Civil War Battlefield
Orders Gone Awry

Also by Donald R. Jermann
Fitz-John Porter, Scapegoat of Second Manassas: The Rise, Fall and Rise of the General Accused of Disobedience (McFarland, 2009)

Antietam: The Lost Order (Pelican, 2006)

Civil War Battlefield Orders Gone Awry

The Written Word and Its Consequences in 13 Engagements

Donald R. Jermann

McFarland & Company, Inc., Publishers
Jefferson, North Carolina, and London

All photographs courtesy Massachusetts Commandery Military Order of the Loyal Legion and the U.S. Army Military History Institute.

LIBRARY OF CONGRESS CATALOGUING-IN-PUBLICATION DATA

Jermann, Donald R.
Civil War battlefield orders gone awry : the written word and its consequences in 13 engagements / Donald R. Jermann.
    p.   cm.
Includes bibliographical references and index.

ISBN 978-0-7864-6949-9
softcover : acid free paper ∞

1. United States — History — Civil War, 1861–1865 — Campaigns. 2. Command of troops — Case studies. 3. Written communication — United States — History —19th century. I. Title.
E470.J55 2012   973.7'3 — dc23   2012012404

BRITISH LIBRARY CATALOGUING DATA ARE AVAILABLE

© 2012 Donald R. Jermann. All rights reserved

No part of this book may be reproduced or transmitted in any form or by any means, electronic or mechanical, including photocopying or recording, or by any information storage and retrieval system, without permission in writing from the publisher.

On the cover: Alfred R. Waud, *Humphrey's Division at the Battle of Fredericksburg*, print from wood engraving 13 ¾" × 20 ½", 1862 (Library of Congress)

Front cover design by Rob Russell

Manufactured in the United States of America

McFarland & Company, Inc., Publishers
Box 611, Jefferson, North Carolina 28640
www.mcfarlandpub.com

To my wife, Florence

Acknowledgments

I would like to thank my daughter, Mary Jermann Amozig,
who provided invaluable assistance in the preparation of this book.

Table of Contents

Acknowledgments vi
Preface 1
Introduction 3

1. Ball's Bluff   13
2. The First Battle of Winchester   25
3. The Battles of Mechanicsville and Gaines Mill   36
4. Malvern Hill   49
5. The Second Battle of Bull Run   62
6. The Maryland Campaign of September 1862   76
7. Perryville   88
8. Fredericksburg   98
9. Vicksburg   110
10. Gettysburg   124
11. Chickamauga   138
12. Spring Hill   152
13. The Battle of Five Forks   162

Appendix: Custer's Last Stand   181
Chapter Notes   197
Bibliography   203
Index   205

Preface

Countless books have been devoted to the subject of how "fighting" determined who won the Civil War. However, few, if any, prior to this, have been devoted to the equally true but less obvious subject of how "writing" determined who won the war.

There was no radio in the Civil War and battles were often of such a magnitude that they could not be controlled by the vision and voice of the on-the-scene commanding general. The normal means of control were for him to send his orders to distant subordinate generals by the written word via courier.

The commanding general may have had supreme competence, brilliant insight, and perfect judgment, but if he was unable to communicate clear, concise, and unambiguous written orders to subordinate generals, all could be for naught.

This book presents 13 examples of how the choice of words influenced the outcome of a battle. The choice of but a few words or a single phrase could be the difference between victory and defeat.

Introduction

The Civil War has been discussed, written about, and examined from just about every possible angle. We have books on leaders, the men in the ranks, battles, supply, armaments, politics, medicine, spies, and just about every other topic one could imagine. However, there is one vital area that has never been written about, and it sometimes determined who won a battle and may have determined who won the war. This is the area—or shall we say the art—of battlefield order writing.

Before we look at examples that demonstrate how the wording of an order could affect the outcome of a battle, let us take a close look at battlefield order writing in the contending armies in the early phases of the war. We will start by examining the consequences of not obeying an order.

As of 1860, the conduct of individuals serving in the U.S. Army was governed by the so-called "Articles of War." Article 9 read as follows:

> Any officer or soldier who shall strike a superior officer, or draw or lift up any weapon, or offer any violence against him, being in the execution of his office, on any pretense, whatsoever, or shall disobey any lawful command of his superior officer, shall suffer death or such other punishment as shall according to the nature of the offense, be inflicted upon him by the sentence of a court-martial.

Thus, disobedience of an order from a senior in the execution of his office was a capital offense that could result in punishment up to and including death. When a junior received a written battlefield order, he was compelled to execute it or to operate within the limits of discretion the order provided. If he received two conflicting orders, he was compelled to execute the last one issued.

The Pre–Civil War Army

Forty-seven years intervened between the close of the War of 1812 in 1814 and the onset of the Civil War in 1861. During this period, the only foreign war that the U.S. was involved in was the Mexican War. It was relatively minor, lasting just 20 months in 1846 and 1847.

Aside from the Mexican War bubble, the army was tiny, ranging between 6,000 and 16,000 men. Its main function was guarding the frontier, and its only combat was with Indians. This combat consisted almost entirely of small unit actions involving 100 or fewer individuals.

The largest organizational unit in the army was the regiment. A regiment was commanded by a colonel and consisted of 12 companies of 100 men each. Each company was commanded by a captain. An entire regiment was not normally based at a single location but was spread out into smaller commands. The culture of the army, then, both before and after the Mexican War, was small unit combat.

The Mexican War itself was small scale compared with what was to follow. The largest single unit was General Scott's army in its campaign from Vera Cruz to Mexico City. This army never exceeded 15,000 men, and was often down to around 10,000.

In the period between the War of 1812 and the Civil War, West Point graduates came to dominate the officer corps. The academy averaged some 40 graduates per year, more than sufficient to meet the army requirements. Many actually served only a short time and then entered civilian life.

The military academy at the time was an engineering school and possibly the most prestigious in the country. A peculiarity was that the top graduates in each class entered the coveted "Corps of Engineers." These engineers were not just concerned with military matters; in fact, they performed all government engineering tasks and could thus spend most, or even all, of their careers engaged in improving the navigation of a river, surveying a route for a railroad or road, building a lighthouse, or even being involved in the construction of the dome of the capitol, which was then in progress.

In sum, the pre–Civil War army officer corps was dominated by West Point graduates, their forte was small unit command and combat, and many of the most talented officers spent most or all of their time completely divorced from military matters.

The Civil War Armies

Just prior to the beginning of the Civil War, the U.S. Army contained 16,376 officers and men, and had no organizational unit larger than a regiment. Within two years, this was to grow into two competing armies of over one and a half million men; one million on the Union side, and 500,000 on the Confederate side.

The two armies were similar in organization, judicial procedures, administration, the handling of secret material, and most other aspects. This is not surprising, as the top hierarchy of both armies was dominated by West Pointers. All eight Confederates reaching the rank of full general were West Pointers, as were 15 of the 18 reaching the next highest rank of lieutenant general. In fact, the president of the Confederacy, Jefferson Davis, was a West Pointer and, in addition, had served as secretary of war in the Pierce cabinet.

The influx of this huge number of new soldiers necessitated a change in organization. Regiments were now gathered into brigades, with usually three to five regiments in a brigade. Brigades were then gathered into divisions, with usually three to five brigades in a division. Divisions were then gathered into corps, with usually three divisions in a corps. Finally, corps were gathered into armies, with two or more corps in an army.

Corps were new to the North American continent, having appeared earlier in European armies. If corps had not existed before, they would now have had to be invented: the very size of the armies necessitated their existence. They first appeared in the Union and Confederate armies de facto, but then became de jure.

To understand the necessity of the creation of corps, let us take as an example Lee's army when it invaded Maryland and Pennsylvania in 1863. It contained about 75,000 men. If all this force advanced by a single road, it would stretch out over 50 miles in length. If attacked in front, it could not even reinforce the front from the rear for two days. Furthermore, it would likely die of starvation as enough supplies could not be brought up along the road of advance to feed it.

The answer was corps. Corps, in effect, were smaller self-contained organizations. Each contained its own infantry, artillery, cavalry, and transport. The idea was for the corps to move and subsist separately by separate routes as an army advanced or retreated, but to unite to fight. The armies were now so large that even when they came together, the corps could be some miles apart. For example, when Lee's army united to fight at the battle of Gettysburg, the road distance between Ewell's corps headquarters on Lee's left and Longstreet's corps headquarters on Lee's right was about five miles. It was all but a necessity now that an army commander give a corps commander some discretion in any orders, as the terrain and enemy as seen by the corps commander may have differed significantly from that seen by the army commander.

There was one area where the contending armies differed. The Union adhered to the tradition that no officer could be senior to George Washington, and George Washington was a lieutenant general. Hence, Union divisions, Union corps, and Union armies were all commanded by major generals. There were only two Union generals that held the rank of lieutenant general: Scott at the beginning of the war, and Grant at the end. Confederate army corps were commanded by lieutenant generals, and Confederate armies by full generals.

The Men Who Commanded the Civil War Armies

The Mexican War was receding in time as the Civil War began. Fourteen classes of West Point graduates had entered the army after the Mexican War.

As of 1861, of the prominent Mexican War generals, General Taylor, was dead. General Scott was general-in-chief of the army, but he was old and infirm and soon to be replaced by the young General McClellan. The only regular army general who had been prominent in the Mexican War and was able to serve for the duration of the Civil War was General Wool. However, Wool, because of his age, was never given a field command.

The command of the armies North and South was turned over to younger men. Almost all of the men destined for top command in both armies were West Point graduates. Some had served in the Mexican War but in almost all cases as junior officers in the ranks of captain and lieutenant. These included both Grant and McClellan.

At the outset of hostilities, many young men were promoted directly from captain to general. The move from captain to general was not merely more of the same. A captain was a specialist. If one were of the infantry, he commanded a company; if of the artillery, a battery; and if of the cavalry, a troop.

A general was likely to command all three elements. Furthermore, most generals had judicial powers. They could convene courts-martial and they could approve or override sentences. Captains could not. Generals operated through staffs. Captains did not. In short, the world of a general was far different from that of a captain.

Many of the young West Point graduates now being advanced to general were not even career officers. They had resigned from the army to pursue civilian careers. These included McClellan, Grant, and Sherman.

The men being advanced now were mostly young. Generals in their 30 had now become the norm, and generals in their 20s were not uncommon.

One often thinks of the grey-bearded Robert E. Lee as being a supremely experienced pre-war officer. In fact, Lee never even rose to regimental command. His highest pre-war post was as deputy commander of the Second Cavalry Regiment, and he spent most of his pre-war career in engineering duties.

The one man in the top hierarchy of the two armies who had commanded more than a regiment in the pre-war period was Confederate General Albert Sydney Johnston. Johnston, in addition to participating in the Texas and Mexican wars, had commanded the 1857 expedition against the Mormons, which involved 2,500 men. Ironically, Johnston was killed in his first Civil War battle, that of Shiloh in 1862. Johnston remained the most senior officer of either side to die in action in the war.

To conclude, in the early phases of the Civil War, men who had never commanded more than a company of 100 men, and some none at all, were now commanding divisions and corps and armies and departments of thousands and tens of thousands of men. It was these men who were writing the battlefield orders.

Battlefield Orders

We will now discuss what a written battlefield order should contain before we cite examples of what they did contain.

First, a battlefield order should specify where the originator is located. Second, and extremely important, the order should contain a time of origin that contains both date and hours and minutes. The rule of good discipline specifies that the order that is the last one issued, not the last one received, is the one that must be obeyed. In the Civil War, when communications were notoriously bad, it was always a distinct possibility that orders would be received out of the sequence in which they were issued. In the modern U.S. Armed Forces, all messages are forever afterward referred to by their originator followed by their date and time of origin.

The next imperative for a written battlefield order is that it indicate who else has a copy. The omission of this simple procedure almost cost the Confederates the war in the fall of 1862. Lee drafted a secret battlefield order for the capture of Harpers Ferry and sent copies to Generals Longstreet, Jackson, Walker, and Hill. When Jackson received his copy, he noted that the order tasked Hill, who was subordinate to him. Not knowing that Hill was already destined to receive a copy, he prepared and sent a copy to Hill. Thus, two copies were sent separately to Hill. As luck would have it, one copy was lost and recovered by the Union. Had McClellan slightly more expeditiously acted upon this information, the Confederate army almost certainly would have been destroyed at Antietam. The circumstances under which one copy of the order to Hill was lost have never been determined. However, it is reasonable to conclude that if two copies had not been sent, one would not have been lost.

The next requirements are that the order be clear, unambiguous, and accurate. In the Civil War, there was almost never the opportunity to gain clarification for an unclear or ambiguous battlefield order. The order must also accurately identify places and not use relative descriptions such as "that hill in front of you."

Most important of all, the order should tell the subordinate what he is required to accomplish, but not how. The recipient can be expected to know his own local situation better than the originator and hence the means of accomplishment should be left to him.

Now we come to the most difficult part in battlefield order writing, the matter of discretion. How much discretion should the commander give to the subordinate?

Let us take an example here of an actual situation, and of the various wordings the commander could have used to task his subordinate. As we will see, the various wordings for the same tasking could result in totally different outcomes.

The situation was as follows. On July 1, 1863, the first day of the battle of Gettysburg, the Confederate army had unified faster than the Union army and thus had a major advantage. Corps commander General Ewell, on Lee's left, was driving the disorganized Union army before him and late in the day had reached the base of Culp's Hill. At the time of Ewell's approach, Culp's Hill was not strongly garrisoned by the Union. Lee, from his position, could see that Culp's Hill was part of a series of ridges and hills that could be easily strongly defended if the Union army was given respite and time. On the other hand, if Culp's Hill were seized, the whole position would become untenable for the Union. Ewell must seize Culp's Hill. Let us look at possible wordings for a battlefield order to Ewell.

1. "I would like you to seize Culp's Hill immediately." This could be construed as a suggestion and Ewell, whose troops were tired and the day was getting late, could decide not to comply.
2. "Seize Culp's Hill immediately." This would be an imperative, giving Ewell no discretion. However, Lee had been previously burned at Malvern Hill, where troops had been wasted trying to comply with an order to seize a hill that the local commanders realized could not be taken.
3. "Seize Culp's Hill if feasible." This would give total discretion to Ewell, who might consider his troops too tired, or that there was not enough daylight left to comply.
4. "The seizure of Culp's Hill before it can be strongly occupied by the enemy is exceedingly important for our future operations. You will attempt its seizure immediately. However, if at any point you consider that to proceed farther would be a waste of your men's lives without any realistic prospects of success, you will cease and report."

In the event, Ewell never attacked Culp's Hill on July 1. By the next day, it was too strongly fortified to be taken and the whole Union line was too strong to be dislodged. Had Ewell taken Culp's Hill on the first, there is an excellent chance that the Confederates would have won the battle of Gettysburg and achieved their independence.

We can thus see that the matter of discretion in a written battle order is all-important. To a large extent, the amount of discretion to be given is dependent upon the commander's assessment of his subordinate. Some generals could be depended upon to do their best to execute their commander's wishes, even though the commander's assessment of the situation varied from their own. Other generals could be considered always to value their own opinions over those of their commanders, and more often than not to substitute their own actions for those of their commander if the degree of discretion permitted.

Order writers gradually learned that they had to provide discretion to cover at least two circumstances: (1) if the situation encountered was different from that envisioned by the commander, and would result in a wastage of men without reasonable prospects of success, the recipient was authorized to use his own judgment; and (2) if the recipient encountered a situation wherein he could achieve greater gain by deviating from his orders, he was authorized to use his own judgment.

Generals' Staffs

Generals, unlike junior officers, had staffs. Staff officers were responsible for various specialties under the general's purview. For example, a general would have a staff officer for supply, one for ordnance, and one for artillery. The most important officer on the staff was the one responsible for administration. On most staffs, he was called the "assistant adjutant general" or AAG. On some staffs, the officer designated for the responsibility was the "chief of staff."

Most correspondence addressed to the command, including post action reports, was normally addressed not to the general, but to the AAG (or COS). Likewise, the AAG drafted and signed most of the outgoing correspondence. This often included written battlefield orders. Here we have another potential problem area. If the general merely told the AAG what he wanted done and the AAG chose the phraseology, the order could contain not the general's true intentions but the AAG's understanding of them. Some more cautious generals dictated vital orders, and some still more cautious generals often wrote them in their own hand. These included Jackson and Grant.

All generals had yet another type assistant. This was the so-called aide-de-camp. All generals had at least one. The aide-de-camp was not a specialist but was an assistant to the general for any and all matters. An aide-de-camp almost always accompanied a general in a combat situation and sometimes at the general's bidding drafted and signed battlefield orders.

Nepotism was rampant during the Civil War and a general's aide-de-camp was frequently a relative.

The Means of Transmission of Battlefield Orders

The circumstances under which written battlefield orders were given to subordinates varied greatly. In some instances, they were simply handed to the junior by the senior after a meeting. In other instances, the participating generals in a pending operation first met and agreed upon an operation, and the confirmatory orders were then sent after they dispersed. However, by far, the most common circumstance involved the transmission of an order to a subordinate without the senior and junior having met to discuss the situation.

There was no radio at the time of the Civil War. There was the telegraph, but its use in transmitting battlefield orders was severely limited. First of all, the telegraph necessitated a wire between the two correspondents. If the wire was cut by the enemy, no communications were possible. If the wire was tapped by the enemy, the enemy could receive anything transmitted via the wire.

Telegraph communications were slow. At the time, they employed only Morse code—dots and dashes. The speed of transmission was limited to the skills of the operators at each end of the line, and the rate of exchange was limited to the capabilities of the least skilled operator. In general, telegraphic communications were allowed only for corps and army level commanders, and then only in those instances where time and terrain permitted the laying of the necessary wire.

The primary and predominant means of disseminating written battlefield orders was the mounted courier. The order was handed to the courier by the AAG or aide in a sealed envelope. Instructions to the courier were written on the outside of the envelope. One instruction was the rate at which the courier could ride. Although these were designed to save unnecessary wear on the horse, they were, in a sense, a priority. "Gallop" obviously indicated that the message was urgent and had to be delivered as soon as possible. Other instructions could be "To be delivered personally to General Hill" or something similar.

It was the duty of the courier to deliver the message, have the recipient sign the envelope as confirmation that the message was delivered, and indicate the time of receipt. The courier was then required to return the signed envelope to his AAG.

Delivery was often much more difficult than one might imagine. First of all, the army could be operating in an area unfamiliar to the courier. Maps were scarce and notoriously inaccurate, and country roads were seldom marked. This could make things difficult enough in daytime; the problem was compounded after dark.

If the courier was operating in enemy terrain and he stopped to ask a local directions, there was always the distinct possibility that he would be purposely misled. The location of the army headquarters was usually in a conspicuous and semi-fixed location and relatively easy to find, but lesser units could be harder to find. If the order was destined for a corps or a division commander, the corps or division could well be in motion and its location only approximately known to the courier.

Last but not least, there was always the danger that the courier would run into the enemy, be captured, and have his message compromised. As a general statement, we may say that the delivery of a message by courier often took longer than one might think, and often much longer. A time of delivery for a message to a recipient within two or three miles might well take three, four, or more hours.

Encrypting the Battlefield Order

We have previously noted that Lee's secret battlefield order for the capture of Harpers Ferry was recovered by the Union, and the compromise almost resulted in the destruction of Lee's army. There were other instances where couriers were captured and their messages compromised. Codes and ciphers were available to both armies in the Civil War. Why did they not routinely use them to encrypt secret orders? Why were almost all written battlefield orders transmitted unencrypted? To understand this, we must take a closer look a the codes and ciphers of the time.

First of all, codes and ciphers are two different things. A code necessitates a codebook.

The codebook contains words and phrases that are commonly used in the drafting of messages. Each word or phrase is represented by a group of letters or digits. Let us say, for example, that each word or phrase is represented by a four-digit group. The drafter prepares the message by looking up the words and phrases he requires to compose the message. He then writes down not the words and phrases but the corresponding four digit groups. The recipient then does the opposite.

Everyone and anyone who might send or receive a message in the code must have a copy of the codebook. If the enemy gained access to any codebook at any time, he could read every message sent. The use of codes in the Civil War was particularly hazardous when a large percentage of the populace on either side sympathized with the enemy.

A cipher, on the other hand, does not require a code book. It involves replacing each individual letter with a different letter by a secret formula. The formula contains a "key," which may be changed each day and for each command. Thus, it is much more secure than a code. For the enemy to read a message, he needs to know both the secret formula and the key for the day of the particular command involved.

Now let us take as an example a simple cipher of a type that was widely known during the Civil War era. We start by laying out the alphabet as indicated below. We then write the key of the day, which in this instance is "FOX," vertically to the left of the alphabet as indicated. We then expand the letters F, O, and X into full alphabets as indicated. We are now ready to encrypt our message.

	A	B	C	D	E	F	G	H	I	J	K	L	M	N	O	P	Q	R	S	T	U	V	W	X	Y	Z
F	G	H	I	J	K	L	M	N	O	P	Q	R	S	T	U	V	W	X	Y	Z	A	B	C	D	E	F
O	P	Q	R	S	T	U	V	W	X	Y	Z	A	B	C	D	E	F	G	H	I	J	K	L	M	N	O
X	Y	Z	A	B	C	D	E	F	G	H	I	J	K	L	M	N	O	P	Q	R	S	T	U	V	W	X

Let us say the message is "Send help!" We encrypt the first letter in the F line by going to our right until we reach the letter S and then moving up to M in the top alphabet line. We do the same for the second letter of our message, but in the O line. We move to our right in the O line until we come to an E, and then go up to the alphabet line where we get a P. We do the same for the third letter in our message, but in the third X line. We move over in the X line until we come to an N, and then go up to the alphabet line where we come to a P. For our fourth letter, we move back to the F line and repeat the process. Thus, the word "Send" would actually be sent as "MPPX."

Let us say the drafter of a message is at army headquarters. He drafts the message, which is one-half of a written page, and hands it to an aide to encipher. The aide sits down at a desk with the draft message, a dispatch pad and a pencil, and begins. The half-page message contains 750 characters. The aide can encipher ten per minute. It thus takes him 75 minutes. When finished, he hands it to a courier for delivery.

The courier finds the intended general recipient and hands it to him. The general and his aide are on horseback in the field. The general sees that the message is enciphered and hands it to his aide to decipher. The aide must hold the dispatch pad, the order and a pencil, and he has only two hands. There is no desk to lay the papers on. He must have someone read the message to him letter by letter, while he deciphers each letter. It takes him two hours while his boss, the general, waits impatiently.

To conclude, ciphers were only sparingly used because, at best, they caused major delays and inconveniences. At worst, if a mistake were made in either enciphering or deciphering, they could not be read at all.

In modern encrypted communications, the encryption is performed "on line." That is, the encryption and decryption process is accomplished in the act of sending. There is no delay and no inconvenience. The sender and receiver handle unencrypted material, but any outsider monitoring the transaction receives the material encrypted.

Recap

During the Civil War, officers could be subjected to the severest penalties for disobeying an order, up to death.

The officers suddenly projected into positions where they were commanding vast bodies of troops had been trained in and were experienced in small unit combat. They were inexperienced in sending written battlefield orders to distant subordinates.

The orders were almost always sent by courier, and due to time constraints, the recipient almost never had the opportunity to request clarification. His actions were dictated by the piece of paper before him.

The situation confronting the recipient of the order was often far different from that visualized by the drafter. Drafters gradually learned that they had to allow the recipient discretion. If not given enough, the recipient may have been forced to waste his men. If given too much, the drafter's wishes may not have been implemented.

Written battlefield order writing was an art, and on this art, the outcome of the war hung in the balance.

1

Ball's Bluff

After the disastrous Union defeat at the battle of Bull Run, young General George B. McClellan was called to Washington to set matters straight. At first he was appointed commanding general of the main Union army in the Washington area. Then he was made general-in-chief of all the armies.

McClellan was an excellent organizer and, if nothing else, thorough. He was determined to shape what was to become the Army of the Potomac into a first-rate, disciplined fighting force, however long it took. He would not be goaded into taking the offensive until he was ready. Thus, while McClellan drilled his army, there was a period of relative quiet on the eastern front.

The Potomac Dividing Line

In the Washington area, McClellan's army occupied both sides of the Potomac River. Then, for some 60 miles up the river to Harpers Ferry, the Potomac divided the contending armies. There was a major Union garrison at Harpers Ferry, and from that point westward, the Union again occupied both banks. For some 60 miles, then, the Potomac was the dividing line between the Union and the Confederacy.

The Potomac was and is a major river, and from the beginning of the country was, for most of its length, the boundary between the states of Maryland and Virginia. The average width of the Potomac for the 60-mile stretch under discussion is about 1,000 feet. It is fast flowing, subject to sudden and wide variations in flow, and is not navigable for commercial shipping upriver beyond Washington.

As of 1861, there were no bridges over the Potomac between Washington and Harpers Ferry and, except for a few fords that could be crossed at low water, the river could be crossed only by ferry.

The Potomac contains numerous islands, the largest of which is Harrison's Island. Harrison's Island is about halfway between Washington and Harpers Ferry. It is about two and a half miles long, but only 350 yards wide. Harrison's Island is at the center of our story about the battle of Ball's Bluff.

The river widens as it flows past Harrison's Island, and the island is closer to the Virginia shore than Maryland shore. The distance between the island and the Virginia shore is about 500 feet. The intervening water is swift flowing and can only be crossed by a strong swimmer — and that minus impediments such as a rifle, cartridge box, haversack, etc.

Map 1
Ball's Bluff

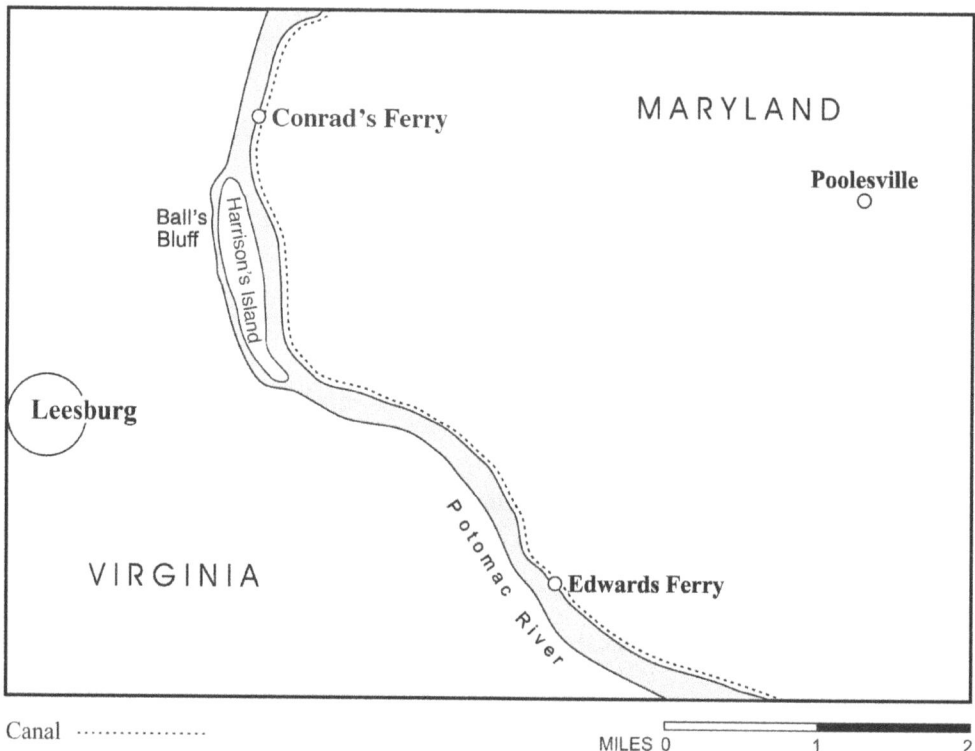

When one crossed from the island to the Virginia side, one encountered a narrow beach and then a steep bluff that ran parallel to the river. The bluff rose to a height of about 100 feet. One could only reach the plateau on top of the bluff by ascending an occasional narrow path. The bluff was called Ball's Bluff, and is the site of our battle.

Approximately two miles south of the bluff was the town of Leesburg, Virginia. Leesburg was the only substantial sized town in the area. It was a major crossroads and the northern terminus of a railroad.

There were two ferry crossings in the area of Harrison's Island. Conrad's Ferry was just to the west of the Island, and Edwards Ferry just to the east (see map 1).

The Chesapeake and Ohio Canal runs parallel to the Potomac on the Union, or Maryland, side. The distance between the river and the canal is usually no more than a stone's throw. The canal tow path runs between the river and the canal.

Although there was and is no commercial shipping other than small ferry boats, on the segment of the river under discussion, there was commercial shipping on the canal. This gave the Union a decided advantage. Although there were no boats for the Confederates to seize, other than the ferries that happened to be on their side of the river, the Union could seize boats from the canal and drag them over the intervening strip of land separating the canal and the river.

The Contending Forces

The main army of the Union in the east was then facing the main Confederate army to the south of Washington. The Confederate army was concentrated in the Centreville area. The secondary armies in the east were facing each other in the Shenandoah Valley with the Union main base at Harpers Ferry and the Confederate main base at Winchester. It is the border between these main concentrations that concerns us here—namely, the Potomac (see map 2).

The Union force entrusted with protecting the gap was called the "Corps of Observation." It was headquartered at Poolesville, Maryland, and was commanded by Brigadier General Charles P. Stone. It contained 6,500 men broken up into three brigades commanded by Brigadier General Goreman, Brigadier General Landers, and Colonel Baker.

The task of the Corps of Observation as given to General Stone by General McClellan was as follows:

> Washington, Aug 11, 1861
> Brig-Gen Charles P. Stone, USV, commanding etc.
>
> General: I have to request that you will proceed with the force placed under your command to the vicinity of Poolesville, and there observe the Potomac River from the Point-of-Rocks to Seneca Mills. You will keep the main body of your force united in a strong position near Poolesville, and observe the dangerous fords with strong pickets, that can dispute the

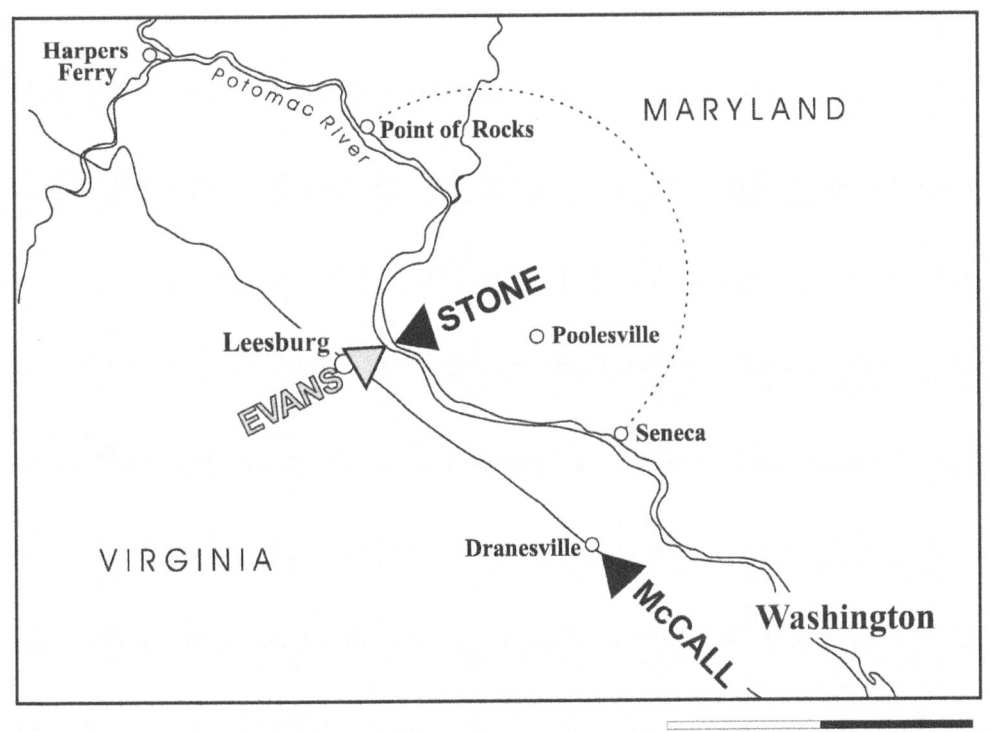

Map 2
Ball's Bluff—The Gap Between the Main Armies

passage until re-enforced. Keep up a constant communication with General Banks pickets near Point-of-Rocks, as well as with those of General McCall and Colonel Smith, until the telegraphic communication is established. Make such arrangements as will enable you, in the event of attack in force, to fall back on General McCall, or enable him to move up to your support at some strong position, which we can hold with the force at our disposal. Should you see the opportunity of capturing or dispersing any small party by crossing the river, you are at liberty to do so, though great discretion is recommended in making such a movement. The general object of your command is to observe and dispute the passage of the river and the advance of the enemy until time is gained to concentrate the reserves of the main force. I leave your operations to your own discretion, in which I have the fullest confidence.

I am sir, very respectfully, your obedient servant.
Geo. B. McClellan
Major-General Commanding[1]

We can see from the above order that the mission of the Corps of Observation was almost entirely defensive, and General Stone's authority to cross the river was extremely circumscribed. It was limited to capitalizing on an opportunity of "capturing or dispersing any small party" and then only by exercising "great discretion." The General McCall that Stone was to fall back upon if in trouble was in command of the western-most unit of the main army located south of Washington (see map 2).

The Confederate counterpart to the Union Corps of Observation was the Confederate brigade of Colonel Nathan George Evans. The brigade consisted of three Mississippi regiments, one Virginia regiment, and three companies of cavalry, for a total of about 1,700 men. The brigade served the same purpose for the Confederates that the Corps of Observation served for the Union. Its purpose was defense. It was to serve as a tripwire to be reinforced if the enemy crossed the river.

As was customary in McClellan's army, the Union estimate of the Confederate strength at Leesburg was greatly exaggerated.

The Opening Moves

The first of the moves that ultimately resulted in the battle of Ball's Bluff came on October 19, 1861. On this date, General McClellan initiated a series of moves that he hoped would induce the Confederates to abandon Leesburg without a fight. This would add to his laurels by affording him a bloodless victory.

On October 19, he ordered General McCall to move his division westward from its position south of Washington and occupy Dranesville, Virginia (see map 2). He would thus appear to be threatening Leesburg from the southeast. McClellan sent the following order to General Stone:

Brigadier-General Stone, Poolesville.

General McClellan desires me to inform you that General McCall occupied Dranesville yesterday and is still there. Will send out heavy reconnaissance to-day in all directions from that point. The general desires that you keep a good lookout upon Leesburg, to see if this movement has the effect to drive them away. Perhaps a slight demonstration on your part would have the effect to move them.
A. V. Colburn
Assistant Adjutant General[2]

In the mind of McClellan, the moves of McCall and Stone were not intended to presage an actual attack, but were merely feints to scare the Confederates out of Leesburg. He did not visualize Stone actually crossing the river, but merely creating the appearance that he was about to do so.[3]

General Stone responded to McClellan's order with the following message:

Poolesville, Oct 20, 1861
Major General McClellan:
    Made a feint of crossing at this place this afternoon, and at the same time started a reconnoitering party towards Leesburg from Harrison's Island. Enemy's pickets retired to intrenchments. Report of reconnoitering party not yet received. I have means of crossing 125 men once in 10 minutes at each of two points. River falling slowly.
    Chas. P. Stone
    Brigadier-General[4]

At this point, McClellan considered the double feint unsuccessful and recalled General McCall from Dranesville to his original position, but failed to so advise General Stone. It is here that the story of the battle of Ball's Bluff really begins, and we will pick up the story with the reconnoitering party that Stone sent across the river from Harrison's Island.

Before we proceed though, let us take a brief look at the leading players.

The Leading Players

The three leading players in our drama were Brigadier-General Stone and Colonel Baker on the Union side, and Colonel Evans on the Confederate side.

Charles Pomeroy Stone graduated seventh out of 41 in the West Point class of 1845. He then was in the thick of the Mexican War, during which he won two brevets.

In 1856, Stone decided that he needed a more lucrative profession and resigned his commission. Up to the time of his resignation, the highest rank that he held was that of first lieutenant. Stone tried various jobs and then settled down as a surveyor for the Mexican government. He held this position for three years until war threatened in the states.

As war approached in January 1861, he was offered and accepted a colonelcy in the Union army. Just eight months later, without having seen combat, he was promoted to brigadier general. On August 12, 1861, Stone, in his first command as general,

Brigadier General Charles P. Stone USV: Victim of Ball's Bluff.

assumed command of the 6,500-man Corps of Observation. Stone's rise was meteoric. Up to January, he had never served in a rank higher than first lieutenant. Now, just eight months later, he was in command of more than a division.

Stone was generally well-liked and respected by both his subordinates and seniors. As a new commanding general, he almost immediately ran into a king-sized problem. Three of his regiments were from Massachusetts, a rabidly anti-slavery state. The colonel of one of Stone's Massachusetts regiments returned a run-away slave that came into the regiment's hands to his master. This was strictly in accordance with the law of the land and Maryland law at the time. However, one or more of his Massachusetts soldiers reported the incident to Governor Anderson of Massachusetts, who chastised the colonel. Stone came to the colonel's defense, pointing out that the act was in accordance with the law, and that the governor should stay out of army business. At this point, the governor was joined by Senator Sumner of Massachusetts, and the two entered into an acrimonious exchange of letters with Stone. Stone now had powerful political enemies as he approached the battle of Ball's Bluff.

Our next leading player was Colonel Edward D. Baker. Baker was no ordinary colonel; in fact, he was one of a kind. At the same time that he was an active duty colonel, he was a sitting United States senator from Oregon.

Baker was a lifelong politician and not a professional soldier. Unlike most of his fellow senior officers, he had never attended West Point or, for that matter, any other military school. Baker's military experience was limited to the Mexican War. At the outset of that war, Baker raised a regiment and became its colonel. In this capacity, he served with distinction throughout the war.

Concerning his attitude toward the Civil War, he was a rabid Unionist who called for immediate, offensive, and unremitting war against the hated Confederacy.

To add to Baker's political clout, he was a close, if not the closest, friend of President Lincoln, with their friendship extending all the way back to the beginning of Lincoln's political career in Springfield, Illinois. Lincoln admired Baker to the extent that he named one of his sons after him.

At the time of our narrative, Baker was in command of one of Stone's three brigades. In looking at the relationship between Stone and his subordinate, Baker, we must consider the following: (1) Baker was a congressman and hence a national figure before Stone even entered West Point; (2) during the Mexican War, Baker was a colonel while Stone was only a second lieutenant;

Colonel Edward D. Baker USV: United States senator and colonel.

(3) at the outset of the Civil War, Baker turned down a major general's commission, a rank higher than Stone's, so that he could keep his status as a senator; (4) Baker was an intimate of Lincoln, the commander-in-chief of all the armed forces, who thus held Stone's future in the palm of his hand; (5) Baker was one of just 44 U.S. senators who had the authority to confirm or reject Stone's appointment as brigadier general; and last (5) Baker was 51 years old while Stone was only 37.

Would this affect Stone's handling of Baker? Of course. Stone was only human.

Confederate Colonel Nathan G. Evans, commanding the brigade at Leesburg, was, like Stone, a West Point graduate. Evans, however, graduated in 1848, three years after Stone. While Stone graduated near the top of his class, Evans graduated near the bottom of his, finishing 36th out of 38. In an era when many of the appointees to West Point had only a rudimentary education, Evans had attended Randolph-Macon College. Consequently, one might have expected that his final placement would be higher.

The war with Mexico was over when Evans graduated, so he lacked Mexican War experience. He did, however, gain considerable combat experience fighting Indians and, in this capacity, had acquired an excellent reputation.

When the Civil War approached and officers in the regular army were choosing sides, Evans was a captain in the Second Cavalry, the regiment of Robert E. Lee. Evans accepted a commission as captain in the Confederate army and was quickly advanced to lieutenant colonel. It was in this rank that Evans was to command a brigade in the first battle of Bull Run. Evans's brigade's performance was outstanding, probably excelling that of all others, including that of "Stonewall" Jackson; and Evans contributed importantly to the Confederate victory.

At this point, Lee, Jackson, and Stuart had not yet ascended as heroes, and it appeared entirely possible that Evans was the one destined for greatness. As a result of Evans's performance, he was promoted to colonel and given the important assignment of the commander of the Seventh Brigade at Leesburg, a billet that would normally be occupied by a brigadier general.

The Battle of Ball's Bluff

As General Stone awaited the return of his scouting party on the 20th, most of his command was split between three locations. These were Edwards Ferry, Harrison's Island, and Conrad's Ferry. Stone himself and General Goreman were at Edwards Ferry. Colonel Baker (fourth in command after Stone, General Goreman, and General Landers) was at Conrad's Ferry, and General Landers was away in Washington on temporary duty.

The scouting party that had been ordered over to Ball's Bluff from Harrison's Island consisted of Captain Philbrick and 20 men. They were directed to cross over, proceed up the bluff, and then follow a little used path toward Leesburg that passed through some woods. Their mission was to surreptitiously locate the enemy, return, and report.

The scouting party finally returned at about 10 P.M. Philbrick reported that they had been undetected and that they approached to within about a mile of Leesburg before encountering anyone. By this time it was dark, and ahead they sighted an encampment of about 30 tents. There apparently were no enemy pickets posted as they were not challenged. At this point they decided to return, and the return was uneventful.

Upon receiving Philbrick's report, Stone decided to exercise the discretion he had been given in McClellan's original order, which read, "Should you see the opportunity of capturing or dispersing any small party by crossing the river you are at liberty to do so."

Immediately upon receiving Philbrick's report, General Stone ordered Colonel Devens, whose regiment was on Harrison's Island, to take five companies of his command across the river, approach the enemy encampment under cover of darkness, and at daybreak attack and destroy it. Colonel Lee was also ordered to cross over from Harrison's Island with 100 men and remain on Ball's Bluff to cover Devens's retirement. The order to Lee and Devens was as follows:

> Headquarters Corps of Observation,
> Poolesville, Oct 20, 1861—10:30 P.M.
> Colonel Devens will land opposite Harrison's Island with five companies of his regiment, and proceed to surprise the camp of the enemy discovered by Captain Philbrick in the direction of Leesburg. The landing and march will be effected with silence and rapidity.
> Col. Lee, Twentieth Massachusetts Volunteers, will immediately after Col. Devens departure, occupy Harrison's Island with four companies of his regiment, and will cause the four-oared boat to be taken across the island to the point of departure of Colonel Devens.
> One company will be thrown across to occupy the heights on the Virginia shore after Col. Devens departure to cover his return.
> Two mountain howitzers will be taken silently up the tow path and carried to the opposite side of the island under the orders of Col Lee.
> Colonel Devens will attack the camp of the enemy at daybreak, and having routed them, will pursue them as far as he deems prudent, and will destroy the camp, if practicable, before returning. He will make all the observations possible on the country; will under all circumstances, keep his command well in hand, and will not sacrifice them to any supposed advantage of rapid pursuit.
> Having accomplished this duty, Colonel Devens will return to his present position, unless he shall see one on the Virginia side, near the river, which he can undoubtedly hold until re-enforced, and one which can be successfully held against largely superior numbers. In such case he will hold on and report.
> Chas P. Stone
> Brigadier General[5]

The above order was well within General Stone's mandate and clear enough right up to the word "unless" in the last paragraph. The final phrase gave Colonel Devens the option of remaining on the Virginia side after he completed his mission. Had it not been for this phrase, there would have been no battle of Ball's Bluff. This last phrase, which was superfluous to the primary mission of Devens destroying the enemy encampment, was to cost General Stone dearly.

Concurrent with the order to Devens, General Stone ordered General Goreman to send a detachment across the river at Edwards Ferry. The purpose of this was twofold. The detachment was to conduct reconnaissance and it was to serve as a distraction from Colonel Devens's crossing. Once across, the Goreman and Devens detachments would be four miles apart and it was not anticipated that they would meet or join in any coordinated attack.

Things started out badly for Colonel Devens and his 300-man detachment and then deteriorated. When his men reached the embarkation point on Harrison's Island at about 1 A.M., they found that there were only three small boats, each capable of carrying about ten men. In consequence, Devens spent between three and four hours getting his men to the Virginia shore. He was not even ready to climb the bluff until about 4 A.M. When Devens finally reached the site of the alleged encampment, he was to receive another surprise. What had appeared to be tents to Philbrick, who had observed in the dark, were actually corn shocks. There was no enemy encampment.

Here Devens turned to the last fateful phrase in his orders. He had met no one, decided that he was not in danger, and that in accordance with the last phrase in his orders, he would stay and request reinforcements. At about 6:30 A.M., he sent Lieutenant Howe to deliver a message of this substance to General Stone at Edwards Ferry. Howe had no sooner left than Confederate cavalry discovered Devens's force, skirmishing began, and Devens began to take casualties.

Howe returned to Colonel Devens at about 10 A.M. with General Stone's response to Devens's request for reinforcements. Stone had agreed to send Devens the remaining six companies of his regiment. This would bring Devens's force in Virginia up to about 650. Stone had also agreed to increase Colonel Lee's force on the bluff that was standing by to cover any retirement of Devens.

Here we can see the seeds of a Union disaster beginning to sprout. When Confederate General Forrest had been asked the secret of his success, he responded, "Get there first with the most men."[6] Forrest expressed the quintessence of good tactics. The one who can concentrate his fire power first at the scene of action wins. In the case at hand with Colonel Devens, there was no physical impediment to the Confederates concentrating at the scene of action. However, the Union could only concentrate in driblets because of the limited shipping between Harrison's Island and Virginia.

While Lieutenant Howe was making his round trip between Colonel Devens and General Stone, the Union troops on Harrison's Island did improve the shipping somewhat, but only somewhat. They had managed to drag another canal boat from the canal to the river, and this one was larger and could carry 60 to 70 troops. Thus, there were four boats available to carry the remainder of Devens's troops over to Virginia; three capable of carrying ten men each, and one capable of carrying 60 to 70 men.

After Howe reported to Colonel Devens that reinforcements were en route, Devens ordered Howe to return once again to Stone and report that Devens was now in contact with the enemy and in action.

Meanwhile, back on the Maryland side, Colonel Baker, Stone's third brigade commander, who had not been involved up to this time, decided to ride over to Edwards Ferry to see General Stone and find out what was going on.

Stone reviewed the situation with Baker and then gave him authority over Stone's entire right flank. This included the forces on Harrison's Island and Conrad's Ferry, consisting of not only Baker's own brigade, but elements of the absent General Landers's brigade. Baker was free to augment Devens on the Virginia side or to withdraw him as he saw fit. Baker requested that Stone reduce his orders to writing. Stone agreed and provided Baker with the following written order:

Headquarters Corps of Observation
Edwards Ferry, Oct 21, 1861
Col E. D. Baker, Commanding Brigade

    Colonel: In case of heavy firing in front of Harrison's Island, you will advance the California regiment of your brigade or retire the regiments under Colonels Lee and Devens upon the Virginia side of the river at your discretion, assuming command on your arrival.

    Chas B. Stone
    Brigadier Gen Commanding[7]

Although the written order was less wide-ranging than the spoken orders, Baker seemed pleased with it and set off toward Harrison's Island to implement it. The time was 9:30 A.M., October 21, 1861.

This proved to be another critical order in the series of events that led to the Union disaster at Ball's Bluff. One of the criteria of good order writing is to tailor the degree of discretion the order gives to what one knows of the characteristics of the recipient. Stone knew, or should have known, that Baker was a firebrand who called for immediate, unremitting, and total war against the rebels. Giving Baker the authority to either withdraw Devens's force or augment it and fight an all out battle on the Virginia side now could have only one outcome. Baker would choose to fight and fight now, and so he did.

Stone later wrote that it would have been possible to withdraw the Devens/Lee force from the Virginia side all morning, as the combat up to that time consisted primarily of skirmishing.[8] Stone's preference, as expressed to Baker in their conversation of that morning, had been to withdraw the Devens/Lee force if it was threatened by a "respectable force."[9]

Baker, after leaving his meeting with Stone, met the oncoming Lieutenant Howe. In their brief conversation, Howe revealed to Baker that he was carrying a message to Stone indicating that Devens was in contact with the enemy. Baker in turn told Howe that he had made the decision to reinforce Devens. Thus, it is evident that Baker had made his decision to transfer his command to the Virginia side within moments of his departure from his meeting with Stone, and without either visiting Devens on the Virginia side to assess the situation, or sending a request to Devens to report his situation. Baker ordered not only his own California regiment to cross over, but the Tammany regiment of Landers as well.

As we have noted, the three ten-man boats available for crossing had now been augmented with a larger boat that could carry 60 to 70 men. However, this new capacity had to cope with new problems. Devens now had wounded to transport back to Harrison's Island. Thus, the boats were delayed on the Virginia side while the wounded and their attendants were loaded. In addition, Baker had cannon, and it took time to embark and debark both them and the horses required to pull them. The shipment of the force across the Potomac was chaotic and slow.

Baker himself arrived at the front at about 2:15 P.M. and took command. By this time, combat was in full swing.

By 4 P.M. things had turned very much in favor of the Confederates. About this time, Colonel Baker was struck by multiple bullets and killed instantly. Colonel Cogswell took command and soon thereafter gave the command to retreat to the river. The first troops to arrive at the river found the three small boats not in sight and quickly swamped the large boat. The retreat quickly degenerated into a rout as the Confederates on top of the bluff fired into the masses of Union troops below, who could find no means of crossing the river. Most threw their rifles and equipment into the river. A few succeeded in swimming across the river but most were drowned, shot, or surrendered.

It was a smashing one-sided Confederate victory. The final casualty total was Confederate 155,[10] Union 921.[11]

Why had the Confederates won, and won so convincingly? First, because of the limited river transportation available to the Union, they were able to concentrate their forces at the scene of conflict more quickly. Second, they exercised superior battlefield management.

This Union tragedy could have been avoided by changes in order writing in two instances. First, General Stone's order to Colonel Devens of 10:30 P.M., October 20, logically ended with the sentence, "Having accomplished this duty, Colonel Devens will return to his present position." However, the order then continued with another phrase giving Devens the option of staying. It was Devens's selection of this option which ultimately proved fatal.

Second, Stone's hastily written order to Baker on the morning of the 21st was in essence a blank check. It gave Baker the option of withdrawing the Virginia force or reinforcing it — without expressing a preference. Because of Baker's known proclivities, there is little to no doubt that he would elect the reinforcement option, even though Stone's own preference was to withdraw the force. Baker should not have been provided a blank check.

Had the orders been different, and had the Virginia force been withdrawn on the morning of the 21st, there would have been no battle of Ball's Bluff. The skirmishing of that morning would have gone down as no more than an insignificant footnote to history, Baker would not have been killed, and Stone would not have ended up in prison (more about this later). Stone was just too generous in giving discretion.

What Happened to Them

The Union debacle at Ball's Bluff had followed the Union fiasco at Bull Run by only 91 days. Twice within three months, the Confederates had roundly trounced a Union force within cannon sound of Washington. The public was up in arms, the press was up in arms, and Congress was up in arms. Someone had to be held responsible. Heads had to roll. Colonel Baker had to all accounts died gallantly at the head of his troops. Hence he, at least for the moment, was beyond criticism. The man in the cross-hairs was Baker's superior, Brigadier General Charles P. Stone.

In December 1861, Congress created the "Joint Committee on the Conduct of the War" to investigate. It called witnesses, some in secret. There was testimony, albeit contradictory and inconclusive, that Stone was disloyal, that he was culpable.

On January 28, 1862, the secretary of war issued an order for Stone's arrest, and he was placed under arrest on February 9. What part, if any, the Joint Committee on the Conduct of the War had in the issuance of the arrest order has never been established.

Stone was initially placed in solitary confinement and, in all, was confined for 189 days — without charges ever being preferred against him. Finally, Congress passed a law that required all those confined to be either charged or released. Stone was then released, but without any explanation, apology, or compensation.

Strangely enough, Stone's rank was never taken from him, and upon release, he was ultimately reassigned. Stone served in relatively unimportant positions until he finally resigned in late 1864.

This, however, was not to be the end of Stone's military career. In 1870, he secured the position of chief-of-staff in the Egyptian army with the rank of Lieutenant General. He then served in this capacity with distinction for 13 years until he retired.

Stone returned to the United States and worked as an engineer until he died in 1887 at the age of 62. He is buried at West Point. Among his final accomplishments as an engineer was the construction of the pedestal for the Statue of Liberty.

Colonel Baker, the second of our threesome, was, of course, dead. However, in death he was now looked upon as a hero, rather than a military incompetent and the true cause of the Union fiasco at Ball's Bluff.

Confederate Colonel Evans was lauded throughout the south for his victory at Ball's Bluff. He received thanks from the Confederate Congress and a gold medal from his native South Carolina. In addition, he was promoted to brigadier general. At the time, it appeared that this was just one more step by Evans on the road to greatness and glory. However, it was to be, in fact, the high point of his career. Evans had not one but two Achilles heels. He had a contentious personality that alienated seniors and subordinates alike, and he was a heavy drinker.

Evans performed well enough as a brigade commander for the next two years, but inevitably ran afoul of a senior in the person of P. G. T. Beauregard. Beauregard lost confidence in Evans and would not entrust him with a field command.

Only a tiny number of senior officers are ever subjected to a court-martial, but Evans was the defendant in two in a single year. The first was for intoxication while on duty, and the second for disobedience of an order. Evans managed to secure an acquittal in both, but then suffered an injury in a riding accident that kept him on the sidelines almost to the end of the war.

Evans secured a position as high school principal after the war, but died prematurely in 1868 at the age of 44.

2

The First Battle of Winchester

After the first battle of Bull Run, the Confederates attempted to prevent a further incursion into Virginia by establishing a defense line across the northern portion of the state. In November 1861, Major General "Stonewall" Jackson was appointed to command what the Confederates termed the "Valley District." This, in effect, was the Shenandoah Valley area which extended across the western part of the state. Jackson's area was thus merely the left flank or the far western part of the defense line.

In March 1862, Union General George B. McClellan transported the main Union army of the east around the Confederate defense line by sea to the York Peninsula. It was from here he intended to move the Union army 70 miles up the peninsula and seize Richmond from the east. The Confederates, of necessity, withdrew their main army of the east from the northern defense line to the peninsula to stand before McClellan and Richmond. Jackson was thus left in the valley unconnected to the main Confederate army.

With this new situation, Jackson's mission changed. His mission was now to tie down as many Union troops as possible to prevent them from reinforcing McClellan on the peninsula. In this, Jackson was outstandingly successful in his fabled "Valley Campaign." In this campaign, he tied down as many as four times the Union troops he possessed, and finally managed to slip away and join the main Confederate army before Richmond, while the Union was still sending reinforcements in the other direction to the valley. It was within this context of Jackson's Valley Campaign that we have our subject, the first battle of Winchester.

The Topography

The first battle of Winchester occurred in the northern part of the Shenandoah Valley, so our description of the valley will be restricted to this area. The valley was bounded by two ridges of the Appalachian Mountains. The eastern ridge was referred to as the Blue Ridge Mountains, and the western ridge as the Allegheny Mountains. The width of the valley averaged about 25 miles, and the Shenandoah River, from which the valley got its name, ran northward along the valley bottom until it emptied into the Potomac (see map 3).

As one moved up the valley toward the Potomac, one came to the city of Harrisonburg. From Harrisonburg northward for the next 40 miles, the valley was divided in two by a slender mountain running down the center. This was Massanutten Mountain. Massanutten was approximately as high as the two parallel ridges bordering the valley.

When the northward-flowing Shenandoah hit the base of Massanutten, it divided into two with one branch flowing on either side. Once past Massanutten, the branches reunited on their flow to the Potomac.

Massanutten ended 50 miles short of the Potomac and, for these final 50 miles, the valley resumed its normal width of about 25 miles.

Approximately halfway down the length of Massanutten, there is a gap that allows one to pass freely from the valley to its left to the valley to its right and vice versa. The small city on the western side of the gap is New Market, and the small city on the east side of the gap is Luray.

Map 3
Winchester — Shenandoah Valley

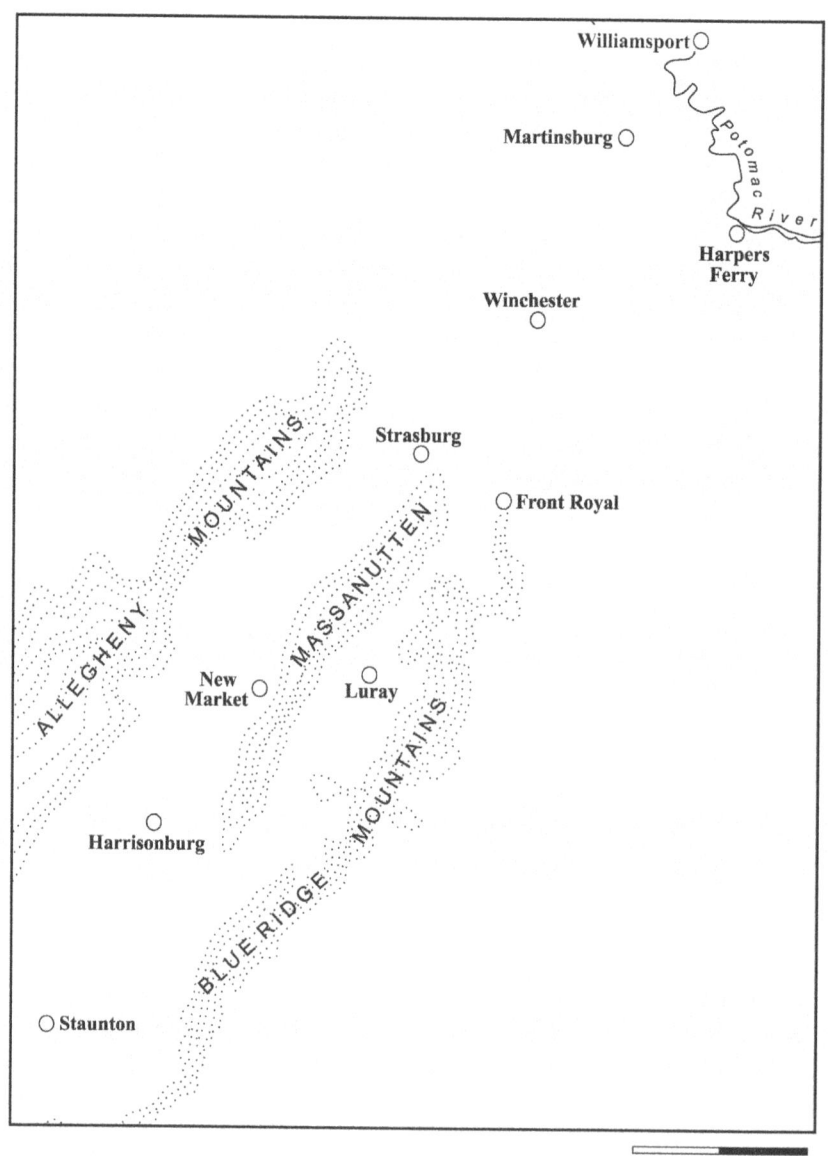

When one reaches the northernmost tip of Massanutten, one again comes to two small cities. The city of Strasburg is at the head of the valley to Massanutten's left (west), and the city of Front Royal is at the head of the valley to Massanutten's right (east). These two cities thus might be likened to corks in a bottle. The military occupation of these two stricture points could effectively stop the flow of traffic either up or down the valley.

There is a 50-mile stretch of open valley from the top of Massanutten to the Potomac. Approximately in its center is the large city of Winchester. Winchester is not only the commercial center of the northern valley, it is its transportation hub. From it, roads radiate to the south to both Strasburg and Front Royal, and to the north to Harpers Ferry and Martinsburg. Harpers Ferry is located on the east side of the valley on the Potomac, and Martinsburg is located on the west side of the valley, not on but near the Potomac.

It can be seen that any Union force that was to stop the movement of Confederate forces from moving up the valley to the Potomac would logically garrison Strasburg and Front Royal, and the Union supply base would be at Winchester. And so the Union was deployed as we begin our narrative.

The Situation on May 20, 1862

The whole Confederate strategy at the time was to draw as many Union troops as possible away from McClellan's army on the peninsula so that Johnston could defeat McClellan before Richmond. The Confederates had achieved considerable success so far in this year. At the last minute, Lincoln decreed that McDowell's corps, one-fourth of McClellan's army, not go to the peninsula with McClellan, but remain so as to occupy a land position between Richmond and Washington.

However, McDowell between Washington and Richmond still provided some advantages to McClellan. The Confederates confronting McClellan before Richmond would now have to look to their flank as well as to their front. The Confederate plan was to ultimately draw McDowell all the way to the far off Shenandoah Valley, where he would have no impact on the Richmond battle.

The first step in this plan was to reinforce Jackson's 8,000 in the valley with another 8,000 under General Ewell. With this augmentation, Jackson could presumably sweep up the valley, cross the Potomac, and then turn east on Washington. This threat would draw McDowell to the valley. Jackson and his army would then slip back to Richmond via the Virginia Central Railroad and, voila, the combined Confederate armies would crush McClellan.

As of May 20, Jackson with his original 8,000 was at New Market, and Ewell with an additional 8,000 was at Luray. Thus, the two were now on the opposite ends of the pass through the center of Massanutten but had not yet joined.

Major General Banks, the Union commander in the valley, was becoming increasingly alarmed. He had only 6,000 men in the valley. About 5,000 were at Strasburg and about 1,000 at Front Royal, and his forward supply base was at Winchester.

On May 22, he sent a telegram of alarm to the War Department outlining his precarious situation and calling for reinforcements. This, of course, is exactly what the Confederates wanted.

Banks's telegram read:

Strasburg, Va May 22, 1862
Sir:

The return of the rebel forces of General Jackson to the valley, after his forced march against Generals Milroy and Schenk, increases my anxiety of the safety of the position I occupy and that of the troops under my command. That he has returned there can be no doubt....

From all the information I can gather — and I do not wish to excite alarm unnecessarily — I am compelled to believe that he meditates attack here. I regard it as certain that he will move as far as New Market, a position which commands the mountain gap ... and enables him also to operate with General Ewell....

Once at New Market they are within 25 miles of Strasburg with a force of not less than

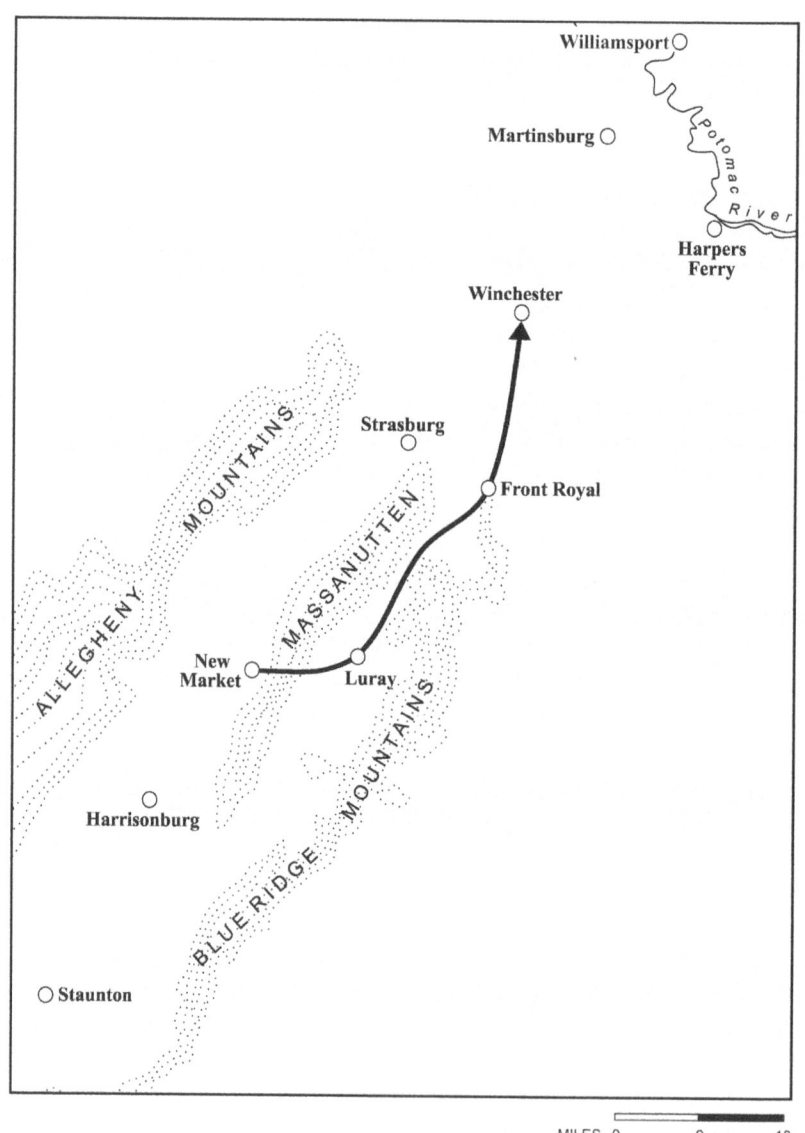

Map 4
Winchester — Jackson's Plan

16,000 men. My available force is between 4,000 and 5,000 infantry, 1,800 cavalry and 16 pieces of artillery.

We are compelled to defend at two points, both equally accessible to the enemy ... my force is insufficient to meet the enemy in such strength as will certainly come.... We greatly need heavier artillery... My infantry should be increased ... the probabilities of danger are so great, that it should be assumed as positive and preparations made to meet it....

N. P. Banks
Major-General Commanding[1]

Banks was right about everything but Jackson's point of attack. Banks believed that Ewell would cross Massanutten, join Jackson, and then the combined force would move against Strasburg. Jackson's plans were other. He planned to cross Massanutten and join Ewell. The combined force would then march up the valley, overwhelm the small union garrison at Front Royal, and then proceed on and seize Winchester in Banks's rear. Banks's force at Strasburg would be doomed to die on the vine (see map 4).

Jackson knew that he must hurry. The farther north he proceeded, the greater the danger. His connection with Richmond was the Virginia Central Railroad, which was entirely in Confederate hands. The Virginia Central hit the valley at Staunton. Staunton was 37 miles down the valley from New Market and over 80 miles down the valley from Winchester. If Federal reinforcements entered the valley between him and Staunton, there was the danger that not only would he be isolated from Richmond, but he would be unable to transfer his army to the peninsula by rail when the time came. The most likely pinch point for the oncoming Union reinforcements to cut off Jackson was Strasburg-Front Royal; Freemont's troops entering the valley on the Strasburg side and McDowell's on the Front Royal side. Jackson had to complete his work north of Strasburg-Front Royal in a hurry and get back south.

The Battle of Front Royal and Winchester

Jackson crossed over Massanutten Mountain to Luray on May 21 and joined his force to that of Ewell. The total force now numbered 16,000. On Thursday, May 22, the entire force moved down the road from Luray toward Front Royal and went into bivouac ten miles short of Front Royal.

Jackson's force moved to the attack at dawn on Friday, May 23. The entire Union force in and around Front Royal was only about 1,000. It was commanded by Colonel Kenly and consisted of nine companies of the First Maryland Infantry, augmented by two 10-pound Parrott cannons manned by 38 artillerymen. In addition, two companies of the Twenty-ninth Pennsylvania Infantry, not under Kenly's command, were nearby guarding a railway bridge and were able to assist in the fight. Two companies of the Fifth N. Y. Cavalry also arrived from Strasburg during the fight to augment Kenly and participate in the fight.

Jackson's advance met no resistance until it reached a point about one and a half miles from Front Royal at about 2 P.M. The ball opened at this point. Although Kenly and his men put up a spirited defense, they were overwhelmed by Jackson's 16,000 and it was all over by dark. Kenly's force was destroyed.

Jackson estimated that he took about 700 prisoners, including 20 officers and Kenly, Kenly's two cannon, and large quantities of commissary and quartermaster stores.[2]

Saturday, May 24, consisted of a race to Winchester. By the early hours of the 24th,

General Banks realized that Jackson's immediate target was not Strasburg but Winchester. He had to get there before Jackson. The road from Strasburg to Winchester was the valley pike, and the distance was approximately 20 miles. Jackson's road from Front Royal to Winchester was approximately 22 miles, so Banks had a slight advantage.

Banks's march started soon after daylight and the whole column was on the march by 9 A.M. The wagon train was initially in the lead, but Banks soon perceived that the danger was in the front and consequently ordered the troops to take the lead.

Jackson, in the meantime, was proceeding on the converging road from Front Royal. Jackson had two cavalry forces under his command. The one attached to his original command was commanded by General Ashby. The other, attached to Ewell's command, was commanded by General Steuart. Jackson used these forces to cross between the roads to attack Banks's moving column.

Ashby hit the column first at Middleton, about 13 miles short of Winchester. Ashby was followed by Jackson's division, while that of Ewell continued on down the Front Royal–Winchester road toward Winchester.

Ashby played havoc with the wagon train, destroying many wagons, but failing to stop the column. Jackson described the situation as follows:

> The turnpike, which had just before teemed with life, presented a most appalling spectacle of carnage and destruction. The road was literally obstructed with the mingled and confused mass of struggling and dying horses and riders. The Federal column was pierced, but what proportion of its strength had passed north toward Winchester I then had no means of knowing.[3]

In fact, most of the column had already passed north to Winchester.

The Federals abandoned many wagons on their rush to Winchester, and these proved a fatal lure to Ashby's cavalrymen. Ashby's men, accustomed to shortages and short rations, were now confronted with a cornucopia of free goodies. Their organization literally dissolved into a mob of plunderers and they were of no further use in the battle of Winchester.

Jackson, observing the disintegration, stated:

> I was pained to see, as I am now to record the fact, that so many of Ashby's command, both cavalry and infantry, forgetful of their high trust as the advance of a pursuing army, deserted their colors, and abandoned themselves to pillage to such an extent as to make it necessary for that gallant officer to discontinue further pursuit.[4]

Banks's column was hit again, a few miles farther down the pike at a place called Newton. This time, the attacker was Ewell's cavalry under General Steuart. Steuart inflicted further damage on the wagon train but, again, failed to stop the column.

Banks reached Winchester after dark ahead of Jackson. Although he had lost much of his supply train, he had lost few men, none of his cannon, and his organization was intact and still full of fight.

His organization took the high ground and awaited Jackson's attack on the morrow. The battle of Winchester occurred on Sunday, May 25. It could have only one outcome. It was Jackson's 16,000 against Banks's less than 5,000.

Banks's men put up a good fight but were ultimately flanked on both sides. Their organization finally broke and it was now a matter of survival. Safety was the other side of the Potomac, and to get there, they must take the road to Martinsburg. A short distance beyond Martinsburg was Williamsport on the Potomac. Williamsport had both a ford and a ferry.

Jackson's tired infantry could not catch the fleeing Federals. Of necessity, he finally went into bivouac as the distance widened.

In the words of Jackson as he watched the fleeing Federals, "Never have I seen an opportunity when it was in the power of cavalry to reap a richer harvest of the fruits of victory."[5]

Jackson had won a handsome victory. He had inflicted over 1,200 casualties versus a loss of 400. He had captured the ground and mountains of supplies. But now, to gather in the full fruits of victory, he needed cavalry. His own division's cavalry, that of Ashby, was useless. He must use that of General Steuart, who was part of Ewell's organization.

Jackson sent for his aide-de-camp, Lieutenant Pendleton. He ordered Pendleton to find Steuart and "order him to move as rapidly as possible and join me on the Martinsburg Turnpike, and carry on the pursuit of the enemy with vigor."[6]

The Finisher Concept

Before we continue our narrative, let us elaborate the concept of "finisher." We might compare war to a boxing match. In a boxing match, one boxer may seriously hurt an opponent, or perhaps knock him down, and then settle for winning the round. After a short rejuvenation period between rounds, the hurt opponent may come back fully recovered and ultimately win the fight. A so called "good finisher" is a boxer who will not let an opponent off the hook once hurt. He will exert every effort to complete the knockout and thus end the fight then and there. There will be afforded no "between the rounds" refreshment period for his opponent to make a comeback.

The great generals of history were all good finishers. For example, Grant was probably not as smart as McClellan and he was certainly not as inspirational a leader, loved by his men. Grant, however, was a finisher, and McClellan was not.

In a battle, once an enemy gives way and is routed and flees the field, it is considered as a victory for his opponent. The winner may have won the ground, inflicted more casualties, and destroyed more property; but unless he "finishes" the enemy, a refreshed enemy will ultimately be back and may then prevail.

The time to secure the fruits of victory and achieve an ultimate victory is after the opponent is routed and fleeing the battlefield in a disorganized fashion. In almost all cases in the Civil War era, the finisher had to be one's cavalry. This was for the simple reason that one's infantry could not catch the fleeing opponent. The cavalry could. It could not only catch, but ride among or ride on ahead. It could burn the enemy's wagons, capture his artillery, kill his draft horses, and ultimately kill or capture the fleeing troops. As we shall see, Jackson won the battle of Winchester, but failed to finish his opponent.

The Pursuit

When we interrupted our narrative, General Jackson had directed his aide-de-camp, Lieutenant Pendleton, to find cavalry commander, General Steuart, and direct him to finish off the fleeing Federals. Here we will turn to Lieutenant Pendleton's own words to see what transpired.

> On Sunday May 25, after the enemy was driven out of Winchester, the pursuit had been carried on with infantry and artillery for some 3 miles toward Martinsburg, when I was directed

by General Jackson to find the cavalry, under Brig. Gen. G. H. Steuart, and send them on at once rapidly, in order that the enemy might be pressed with vigor. This was about 10 o'clock in the morning. I rode rapidly to Winchester, and failing to ascertain the whereabouts of the cavalry by inquiry, I determined to go to Major-General Ewell, on the east of Winchester, under whose immediate command General Steuart was acting.

I found the cavalry some 2½ miles from Winchester, on the Berryville road, with the men dismounted, and the horses grazing quietly in a clover field. Not seeing General Steuart, I gave the order direct to the Colonels of the regiments to mount and go rapidly forward to join General Jackson on the Martinsburg Turnpike. Colonel Flournoy, Sixth Virginia Cavalry, the senior colonel, requested me to ride on and overtake General Steuart and communicate the order to him, as he had directed them to await him there. Going some half a mile farther, I overtook General Steuart, and directed, by General Jackson's order, to move as rapidly as possible to join him on the Martinsburg Turnpike and carry on the pursuit of the enemy with vigor. He replied that he was under command of General Ewell and that the order must come through him. I answered that the order from General Jackson to join him [General Jackson] was peremptory and immediate, and that I would go forward and inform General Ewell that the cavalry was sent off. I left him, and went on some 2 miles and communicated with General Ewell, who seemed surprised that General Steuart had not gone immediately upon receipt of the order.

Returning about a mile, I found that, instead of taking the cavalry, General Steuart had ridden slowly after me toward General Ewell. I told him I had seen General Ewell and brought the order from him for the cavalry to go to General Jackson. This satisfied him. He rode back to his command, had them mounted and formed, and moved off toward Stephenson's Depot.[7]

And so, one hour was wasted and Jackson saw the fruits of his victory slip through his fingers. It is true he had won a victory. He had driven the Yankees from the valley back across the Potomac to Maryland. He had inflicted over 1,200 casualties wherein he had taken only about 400. He had captured mountains of supplies. And yet, the victory was far short of what it could have been. Banks and most of his troops reached safety and lived to fight another day. They would be refurbished and refreshed for the next round.

What Went Wrong?

The pursuit failed to achieve its desired effect because it was delayed over an hour when Colonel Flournoy and General Steuart declined to act instantly upon Jackson's order. Flournoy kicked the matter up to Steuart and then Steuart kicked it up to Ewell.

Colonel Flournoy, although a political officer, almost certainly knew that Jackson had the authority to order him directly if he so desired, without going through Ewell. Even if Flournoy did not know this, General Steuart certainly did. Steuart, unlike Flournoy, was a career officer. He both graduated from West Point and had 13 years experience in the regular army before volunteering for the Confederate army.

Let us look more closely at exactly what transpired. Lieutenant Pendleton was part of Jackson's original command. Both Flournoy and Steuart were part of Ewell's command. Ewell's command had only physically fused with Jackson's two days before the battle. It is entirely possible that at least Flournoy, and possibly Steuart, not only did not know Pendleton, but did not recognize him by sight. At least Flournoy's first encounter with him may well have occurred when Pendleton appeared before him on the 25th of May.

2. The First Battle of Winchester

Lieutenant General "Stonewall" Jackson CSA: Master of the Valley.

Pendleton, at the time, was a first lieutenant. He had no military experience before receiving a commission in the Confederate army, and this was just his 53rd week of service. He was, at the time he rode up to Colonel Flournoy, 21 years old — and looked younger. Flournoy was 50 years old. He was not only a colonel, but had been a United States congressman and candidate for the governorship of Virginia. The gangling youth before him did not hand him an envelope containing an order from Jackson. Rather, he conveyed a spoken order from Jackson. The lieutenant was ordering the colonel to do something contrary to what General Steuart, just minutes before, had told him to do. The young man was ordering Flournoy to saddle up the two regiments immediately and proceed to join Jackson on the Martinsburg Pike. Is there any wonder that Flournoy, a lifelong politician, was wary? That he wanted to play it safe? He quite logically palmed Pendleton off to Steuart, who was not far off.

When Pendleton reached Steuart, he received the same reaction. The cautious Steuart palmed him off to Ewell.

When General Lee sent a verbal order to a subordinate general, or an order that required explanation, elucidation, or discussion, he typically sent his chief-of-staff, Colonel Chilton. Subordinate generals would not only know who Chilton was, but recognize that he was empowered to speak for his commander. Lee thus avoided the situation of an unknown lieutenant ordering a general about.

Jackson was known to be a man who would not tolerate the slightest disobedience to one of his orders. Just weeks earlier, he had relieved General Garnett of command and ordered him court-martialed for what Jackson considered a failure to carry out an order. Why then did Jackson tolerate Flournoy's and Steuart's actions? Jackson contented himself with appending Pendleton's version of what happened to his official post battle report and took no further action. Jackson's lack of action can probably be attributed to his assessment that, if the case went to court-martial, he would not win. The case would come down to what the lieutenant said and what the colonel and general said. The highly regarded Flournoy and Steuart would most likely be exonerated.

Now let us assume that the matter of Jackson's order was handled more formally, as it should have been. The order would have been in writing and signed either by Jackson or his AAG. It would then have been placed in a sealed envelope and handed to a courier, in this case Pendleton, for delivery. The order might have read approximately as follows:

> May 25 10:15 A.M.
> Headquarters on the Martinsburg Turnpike
>   Brig. Gen. Steuart:
>   1) Immediately upon receipt of this order, you will take all of the cavalry forces readily available and pursue, overtake, and destroy the Federal forces fleeing on the Martinsburg Turnpike toward Martinsburg.
>   2) A copy of this order is being provided to Major General Ewell.
>   T. J. Jackson
>   Major-General Commanding

Upon finding Colonel Flournoy, Pendleton would have handed Flournoy the envelope. Flournoy, in accordance with established procedures, would have been required to sign the envelope and indicate in writing on the envelope the time he received it. Pendleton, in turn, would have returned the signed envelope to Jackson.

Under the circumstances, it would have been perfectly clear to Flournoy that, if he did not start out in compliance close to the time of the receipt indicated on the envelope, he would be looking a court-martial and conviction straight in the eye. Had Flournoy complied with the order, Jackson would have successfully finished off Banks.

The Leading Players

Before leaving the battle of Winchester, let us take a brief look at the leading players. "Stonewall" Jackson is so well known that we need say little about him. His "Valley Campaign," of which the battle of Winchester is a part, is universally considered a military masterpiece. However, immediately afterward, upon transferring to the York Peninsula, his performance was pedestrian at best. Why the difference? Most historians try to explain it away by saying that he was tired, fatigued, ill, etc.

The difference in performance can be, at least in part, attributed to a more mundane factor. Jackson's home was in the valley. He lived and worked in Lexington, a small city in the center of the valley. His and his

Major General Nathaniel P. Banks USV: The consummate political general.

wife's favorite recreation was taking carriage rides in the valley. In addition, at the outset of the war, Jackson was placed in command of Harpers Ferry, a city at the north end of the valley. In short, Jackson knew the valley like the back of his hand. While his opponents labored over inaccurate and incomplete maps and sometimes no maps, he knew every peak and valley, every river and stream. He knew where every road led and its condition. In short, while in the valley, Jackson had an advantage over his opponents.

Jackson's opponent was Major General Nathaniel Banks. Banks was the quintessential political general. He had been speaker of the Massachusetts House, governor of Massachusetts, and speaker of the U.S. House of Representatives. When appointed major general, he had no prior military experience. He was appointed not only major general, but one of the senior generals in the entire Union army. His date of rank was just two days after that of McClellan and over four months before that of the general-in-chief of all the armies, Halleck. Banks immediately took to military pomp and circumstance, and was always meticulously uniformed. He was a prominent figure in the war from beginning to end, and when not in the headlines was never far away.

Historians generally rate Banks as a military incompetent. But was he? Banks squared off against Jackson twice in the Civil War as opposing commanding generals. In the instance we have just described, the battle of Winchester, he managed to salvage most of his force when confronted by odds of 16 to five. One would have to rate this performance as "not bad."

Banks's second encounter with Jackson was at the battle of Cedar Mountain on August 9, 1862. In this instance, Jackson managed to eke out a hard fought victory while he outnumbered Banks two to one. One would have to rate Banks's performance as "pretty good." Just perhaps, history has been a bit too kind to Jackson and a bit too hard on Banks.

Our next, and last, leading player is Alexander Swift (Sandie) Pendleton, Jackson's young aide-de-camp. Pendleton was far better educated than most in his time. While the average man of the time did not have a 12th-grade education, Pendleton was not only a college graduate, but attended graduate school. Like most aides-de-camp, Pendleton had been associated with his general before the war. Nepotism was rampant during the Civil War, and numerous generals had sons, nephews, relatives, and relatives of relatives or friends as their aides. Pendleton was not a relative of Jackson but knew him from Jackson's days as professor at VMI. Pendleton, at the time, was a student at George Washington College, which campus adjoined that of VMI. Both Jackson and Pendleton belonged to the same literary society; Jackson a professor, Pendleton a student. This relationship suited them for their future relationship as general and aide. Pendleton shared Jackson's religious ardor and the two were inseparable, Jackson depending ever more on Pendleton. Pendleton was considered the staff officer par excellence and was, soon after the battle of Winchester, promoted to Jackson's AAG, a post he retained until Jackson's death in May 1863. Pendleton was not to survive the war, but was wounded at the battle of Fisher's Hill and died on September 23, 1864, just short of his 25th birthday.

3

The Battles of Mechanicsville and Gaines Mill

Our next case is the Confederate battle order for the battles of Mechanicsville and Gaines Mill. In order to understand these, we must place them within the context of the campaign in which they occurred. The campaign was General McClellan's so-called Peninsular Campaign.

President Lincoln placed the young George McClellan in charge of the troops in the Washington area after the disaster of the first battle of Bull Run. McClellan, an excellent organizer, shaped the forces into what became known as the formidable Army of the Potomac. He then came up with a plan. To end the rebellion, it would be necessary to destroy the main Confederate army in the east that was then located in northern Virginia at Bull Run.

The Confederate capital at Richmond was of greater importance to the Confederacy than merely being its political center. It was also its industrial center and its main producer of armaments. The Confederate army in Virginia could be expected to defend it to the death. Thus, to take Richmond and destroy the Confederate army were really one and the same task.

To capture Richmond, the average layman would proceed south, overland from Washington to Richmond, a distance of 100 miles. This, however, posed two major problems. First, it would entail an ever longer supply line through hostile territory to supply the fighting forces at the front. This supply line had to be protected. Thus, as they say, the advancing army would continue to sacrifice teeth for tail. The fighting force at the front would be ever more attenuated. Second, almost all the rivers in Virginia flowed from west to east. Each of these provided a favorable defense line, and each had to be crossed.

McClellan proposed to obviate these problems by capitalizing on Union control of the seas. At sea, the Union could transport men and material cheaply and efficiently, without the slightest fear of Confederate molestation.

At the outset of the war, the Union retained control of an outpost deep in Confederate territory. This was Fort Monroe, which was located in southern Virginia at the tip of the peninsula between and the York and James Rivers (see map 5). This peninsula led directly into Richmond. McClellan proposed shipping his army of some 120,000 men by sea directly from the Washington area to Fort Monroe. Once at Fort Monroe, the army would still be some 75 miles from Richmond so, superficially, it would appear that McClellan's plan offered little gain. However, that was not all McClellan proposed to do by sea. Just 20 miles down

Map 5
Mechanicsville — McClellan's Plan (Phase 1)

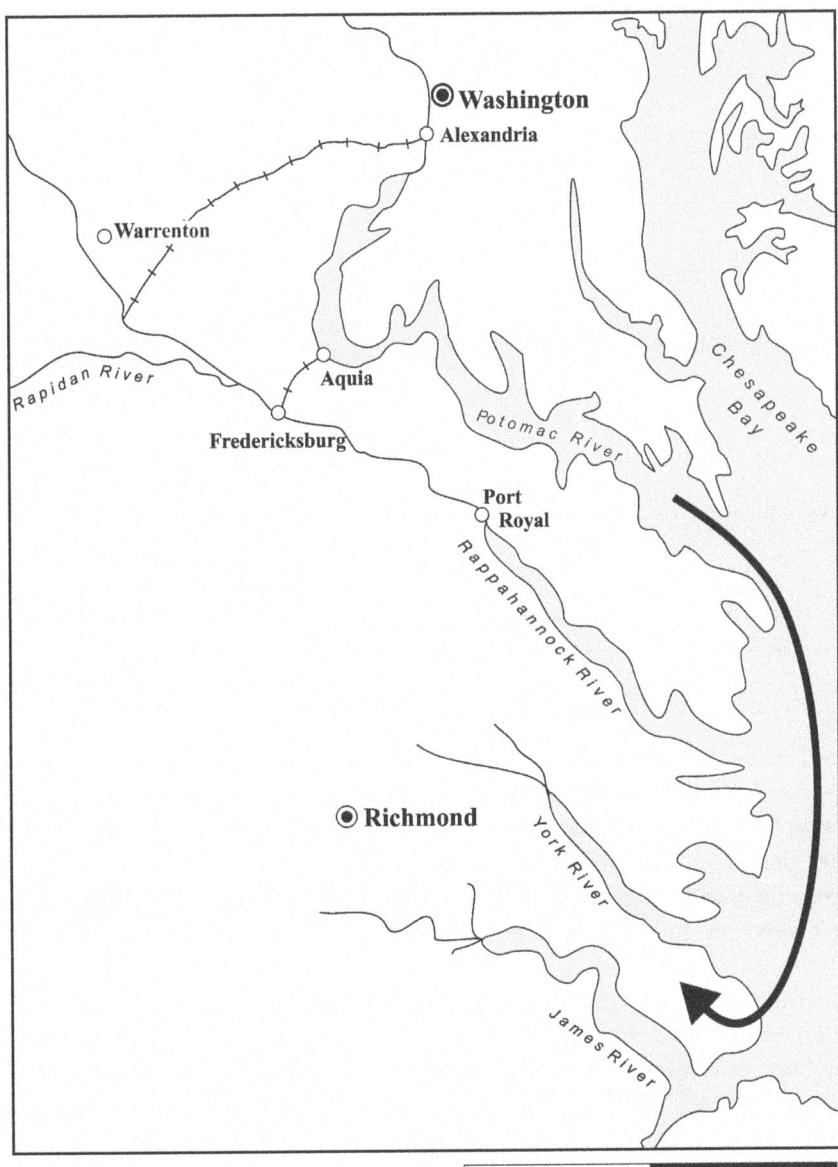

the peninsula from Fort Monroe, the wide York River momentarily narrowed into a channel just 1,000 yards wide before widening again. This pinch point was guarded by a Confederate fort at Yorktown on the peninsula, and another fort on the opposing side on the mainland at Gloucester. If McClellan could seize these forts, the wide York would be open to Union shipping. In this event, the Union forces could move by sea all the way up the York to a place called White House, which was less than 30 miles to Richmond. To put the icing on the cake, White House was on a rail line that led directly into Richmond (see map 6).

**Map 6
Mechanicsville — McClellan's Plan (Phase 2)**

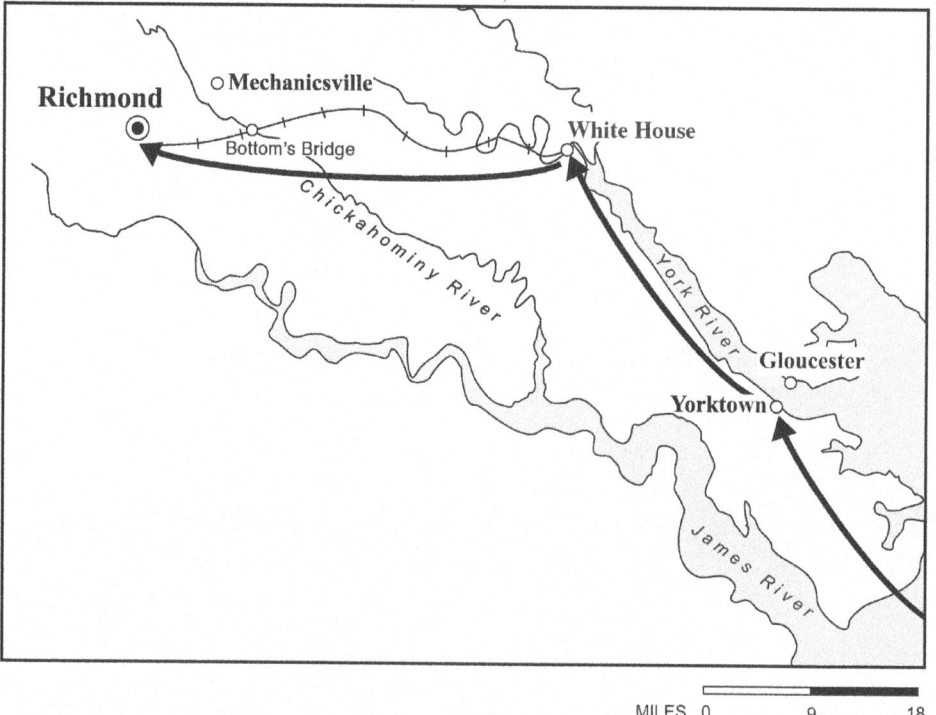

McClellan's plan called for him, once ensconced at Fort Monroe, to seize the forts at Yorktown and Gloucester, open the York to Union shipping, seize White House and establish it as his supply base, and then move down the rail line to Richmond. McClellan anticipated that, once he seized White House, the Confederates would move all the railway rolling stock down the line to keep it out of his reach, but he even provided for this. He ordered that five locomotives and 80 freight cars be shipped to him once he was established at White House.

For a time McClellan's plan progressed swimmingly. McClellan himself arrived at Fort Monroe on April 2, 1862. On May 5, Yorktown was his and the York River bottleneck was open. By May 10, White House was his and he had a "safe" base of operations less than 30 miles from Richmond. He then began to push the Confederates down the railroad toward Richmond.

Before proceeding, let us look at the contending forces at this time. McClellan's army consisted of five corps plus cavalry. The corps were commanded as follows: Second by Sumner, Third by Heintzelman, Fourth by Keyes, Fifth by Porter, and Sixth by Franklin. The cavalry was commanded by Stoneman. The Confederate army was commanded by Joseph Johnston. Up to this time it had not been formally broken up into corps, but consisted of a number of commands, some consisting of a single division and others of more than one division. These were the commands of Magruder, Huger, Longstreet, A. P. Hill, and D. H. Hill. The cavalry was commanded by Stuart. At this point the Union army totaled about 120,000, and the Confederate army about half that number.

Things went well for McClellan until he pushed down as far as the Chickahominy River.

THE CHICKAHOMINY PROBLEM

The Chickahominy starts northwest of Richmond and runs diagonally across the peninsula until emptying into the James River 45 miles east of Richmond (see map 6). As McClellan studied his maps in Washington before beginning the campaign, the Chickahominy appeared to pose no problem. It appeared to be a small, shallow, slow-flowing river, only about 40 feet wide in its upper reaches, that could be easily forded anywhere. However, the maps concealed a nasty secret that McClellan was soon to learn. The upper reaches proceeded through a swampy lowland, and after any period of rain, the river widened to one-half to one mile of impassable swamp. Furthermore, the approaches to the bridges had to contain long sections of corduroy that tended to wash away when the river flooded. Richmond lay to the south of the Chickahominy, and if McClellan was going to take Richmond he had to cross the river.

Now every good general knows that the last thing you want is to have the halves of your army divided by an impassable river while your opponent has his entire army on one side. McClellan and Johnston, both being good generals, recognized this fact.

By May 20, McClellan's advance guard reached the Chickahominy at Bottom's Bridge (see map 6) and secured a beachhead on the south side. By this time, the entire Confederate army had retired to the south side. McClellan's army then began crossing. By the last days in May, the corps of Heintzelman and Keyes were across, while those of Sumner, Porter, and Franklin were still north of the river. Then the rains came — and came and came. Most of the bridges flooded out, and those that remained were becoming ever more questionable.

General Johnston, seizing his chance, attacked the isolated corps of Heintzelman and Keys on May 31 in the battle called "Seven Pines." McClellan's luck held. Two things saved McClellan. The Confederate attack was poorly coordinated, and Sumner's corps was able to cross a shaky bridge to come to Heintzelman's and Keyes's aid. It was, however, a close call and shook McClellan. Even before this close call, McClellan had come to the conclusion that he would be better off if he transferred his base of operations at White House on the York side of the peninsula to the James River side.[1] If he were to do this, Richmond, his base of operations, and his entire army could be kept on the south side of the Chickahominy. It would never have to be divided. Although this was now McClellan's desire, he had a problem that prevented its implementation.

Before we go into McClellan's problem, we must note a happening at the battle of Seven Pines that had a profound effect on the future of the war. The cautious Confederate army commander, Joseph Johnston, was seriously wounded and was replaced by the aggressive Robert E. Lee.

THE MCDOWELL PROBLEM

When McClellan initially revealed his plan to attack Richmond from the York-James Peninsula, he immediately encountered a problem. Lincoln didn't like the plan. Lincoln was fearful for the safety of Washington. McClellan's plan left no large force between Washington and the Confederate army defending Richmond. Lincoln feared that the Confederates might decide to give up Richmond and march north and seize Washington. Consequently, Lincoln, at the last minute, ordered one of McClellan's corps to remain behind to take up a position on the Rappahannock River halfway between Washington and Richmond. This was the First Corps, commanded by Irvin McDowell. This became and remained a sore point with McClellan.

By May 18, McClellan had already decided to move his supply base from White House to the James. However, on that date he received a telegram from the secretary of war advising him that McDowell would advance from Fredericksburg and directing him to extend the right of the Army of the Potomac to the north of Richmond in order to establish communications with McDowell. The same telegram also required him to supply McDowell from his depots at White House.[2] This and subsequent directions and advisories from Washington promising the junction of McDowell's and McClellan's forces required McClellan to retain a presence north of the Chickahominy and the supply base at White House far longer than he wanted to. As it happened, there never was a junction between McDowell and McClellan.

The Jackson Factor

From the outset of the war, there were two distinct theatres of war in Virginia where separate armies contended. The main theater where the main armies contended was the great coastal plain, which extended from the ocean to approximately 150 miles to the west to the Appalachian Mountains. The secondary theater was the Shenandoah Valley in the Appalachians, which ran roughly parallel to the coast. The Shenandoah was valuable both materially and strategically. It was a rich fertile farmland that was considered the bread basket of Virginia. In addition, it provided an excellent avenue for an invasion of the North, as well as an avenue to cut the rail connections between the east and west portions of the Union. In short, it was important to both sides, and the opposing armies fought up and down the valley throughout the war.

In May 1862 when Lee took command, the Confederate Valley Army was commanded by the redoubtable "Stonewall" Jackson, perhaps the best general in the Confederacy.

The perceptive and aggressive Lee was undoubtedly aware that the greatest Confederate victory of the war to date, that of Bull Run, was achieved by the Confederate Army of the Valley slipping away from its opponent and joining the main army on the plain. The united armies then did what they could not do separately — they routed the main Union army. This worked once. Why not try it again?

On June 11, Lee sent Jackson a message calling upon him to secretly join Lee's army near Richmond.[3] Jackson's troops were to come to Ashland, using the Virginia Central Railroad to the degree possible (see map 7). To enhance the secrecy surrounding Jackson's withdrawal, Lee sent Jackson reinforcements, thus implying to the Union that Jackson was about to take the offensive in the valley. The ruse was so successful that, as Jackson was heading for Richmond, McDowell was heading toward the valley. Lee's message also called for Jackson to come personally to Richmond for discussions prior to the arrival of his troops.

Jackson arrived at Lee's headquarters in Richmond on June 23 at about noon. He had covered the final 52 miles of his journey by riding relay horses since 1 A.M. that date.[4] To

Map 7
Mechanicsville — The Virginia Central Railroad

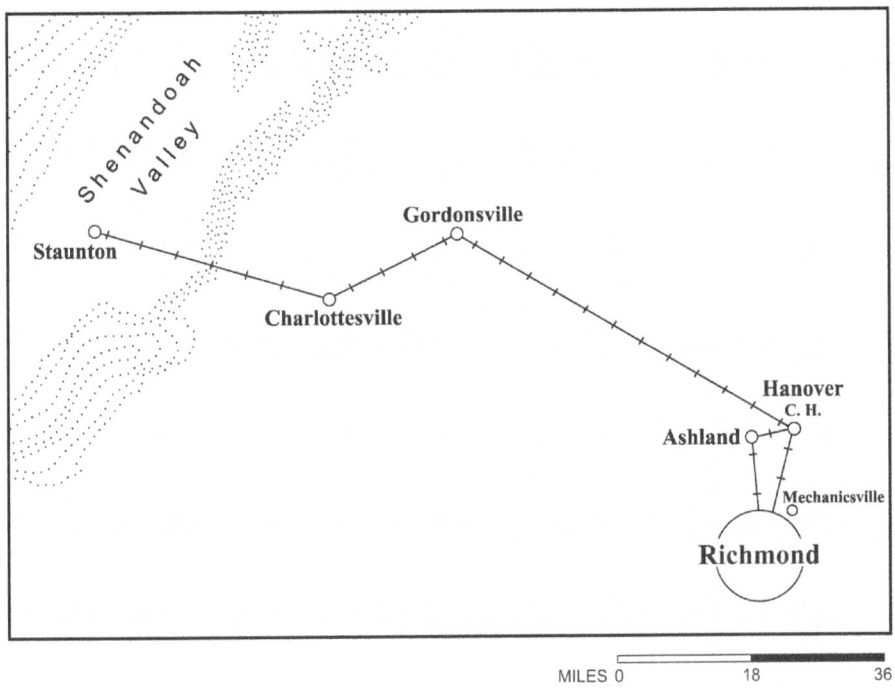

avoid being recognized in the Richmond area, Jackson traveled without insignia and without escort. Lee had also invited three other generals whose commands were to take part in his forthcoming offensive. These were Generals Longstreet, A. P. Hill, and D. H. Hill. The plans had been so secretive that D. H. Hill was amazed to see Jackson at the meeting. He thought him far off in the valley.

The situation confronting Lee was as follows: His entire army was south of the Chickahominy (see map 8). Four of the five corps of the Union army were south of the Chickahominy. The fifth, Porter's, was north of the river still hoping for a meeting with McDowell and guarding White House and the rail line supplying the corps to the south. Porter's corps consisted of about 30,000 men.

Lee's target was Porter. To get at Porter, Lee was going to use the forces of Longstreet, A. P. Hill, and D. H. Hill of his army, plus Jackson's command. The combined forces of the two Hills and Longstreet was about 30,000. Jackson had 30,000 more. Lee thus planned to attack Porter's 30,000 men with 60,000. Lee planned to leave the forces of Magruder and Huger dug in south of the river to confront the four Union corps that were south of the river.

To get at Porter, the Hills-Longstreet force had to get across the river. Ashland, where the Jackson force would arrive, was already north of the river. Lee planned to cross via the three westernmost bridges: Half Sink, Meadow, and the bridge at Mechanicsville. In all three instances, the Union forces of Porter were on the north side and those of Lee on the south side. However, the bridges were intact. The reason they were intact was because in this area of the river there was a high bank on the Confederate side that allowed them to mount cannon to sweep the approaches on the Union side.

Porter was no fool and knew that if he was to be attacked it would be via these bridges. His main force was concentrated in the Mechanicsville area (see map 8). If attacked, he had decided not to make his main defense at the bridges, but to withdraw downriver to a point just beyond Mechanicsville where Beaver Dam Creek entered the Chickahominy. Beaver Dam Creek ran roughly perpendicular to the Chickahominy. The far bank of the creek was much higher than the near bank and it was here he would concentrate his forces. The approach area to Beaver Dam Creek on the side the attack must come was wide open, without any potential cover for the attacking forces. Thus, the Beaver Dam defense line would make a most formidable defensive line and the defenders could easily turn back a force much larger than themselves.

Map 8
Mechanicsville

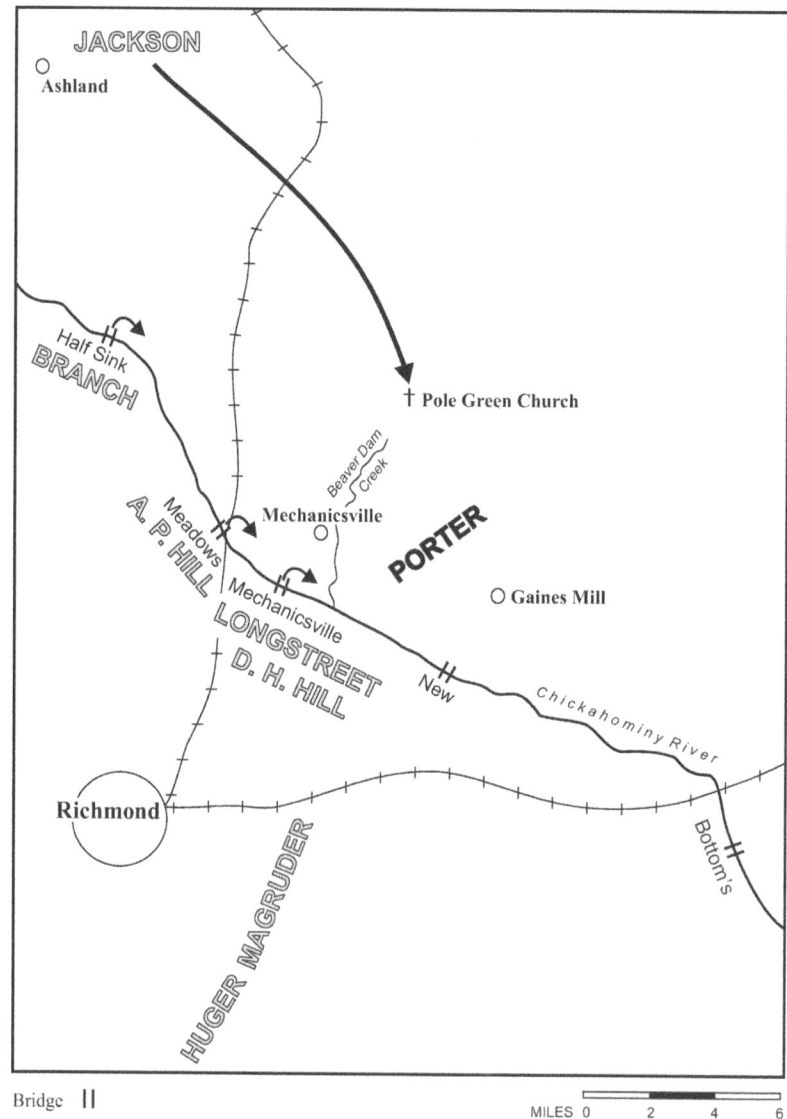

Lee, in planning his campaign against Porter, fully understood that this was the barrier the Hills-Longstreet force would ultimately encounter once it was across the river. Lee's ace in the hole would be Jackson's force, which, once it arrived at Ashland, would already be on Porter's side of the river and be north of the northern flank of the Beaver Dam line. Lee's basic plan was as follows: the Longstreet-Hill group would force passage at the three bridges and push Porter's defenders back down to their Beaver Dam defense line. While the Hills-Longstreet force was approaching the defense line, the Jackson force would be circling around the northern flank of the line to attack it in the rear (see map 8). It sounded great. It appeared to have all the ingredients for a smashing victory. Once the Porter force was eliminated, Lee could capture the rail supply line to the Union forces south of the river and seize the supply base at White House.

As the meeting of the generals progressed, Longstreet pointed out that, inasmuch as Jackson had the greatest distance to cover, the movements of the other forces should be keyed on his. He asked Jackson at what time his forces could be in position to begin the attack. Jackson replied, "Daylight on the 26th."[5] Longstreet then said: "You will encounter Federal cavalry and roads blocked by felled timber, if nothing more formidable, aught you not to give yourself more time."[6] The meeting broke up at dark and the generals returned to their commands. This was Jackson's last opportunity to discuss the plan with Lee or the other generals in person.

The plan was reduced to writing and disseminated to the players on the 24th. It was signed by Lee's chief-of-staff, Colonel Chilton. It is not known if Lee dictated the plan to Chilton, or if Chilton merely wrote up his understanding of the plan.

The written plan was as follows:

General Orders
No. 75
HDQTRS Army of Northern Virginia
June 24 1862

    I. General Jackson's command will proceed tomorrow from Ashland toward the Slash Church and encamp at some convenient point west of the Central Railroad. Branch's brigade of A. P. Hill's division, will also to-morrow evening take position on the Chickahominy near Half Sink. At 3 o'clock Thursday morning, 26th the instant, General Jackson will advance on the road leading to Pole Green Church, communicating his march to General Branch, who will immediately cross the Chickahominy and take the road leading to Mechanicsville. As soon as the movements of these columns are discovered, General A. P. Hill, with the rest of his division, will cross the Chickahominy near Meadow Bridge and move direct upon Mechanicsville. To aid his advance, the heavy batteries on the Chickahominy will at the proper time open upon the batteries at Mechanicsville. The enemy being driven from Mechanicsville, and the passage across the bridge opened, General Longstreet and his division and that of General D. H. Hill, will cross the Chickahominy at or near that point, General D. H. Hill moving to the support of General Jackson and General Longstreet supporting General A. P. Hill. The four divisions, keeping in communication with each other and moving en echelon on separate roads, if practicable, the left division in advance, with skirmishers and sharpshooters extending their front, will sweep down the Chickahominy and endeavor to drive the enemy from his position above New Bridge, General Jackson bearing well to his left, turning Beaver Dam Creek and taking the direction toward Cold Harbor. They will then press forward toward the York River Railroad, closing upon the enemy's rear and forcing him down the Chickahominy. Any advance of the enemy toward Richmond will be prevented by vigorously following his rear and crippling and arresting his progress.

    II. The divisions under Generals Huger and Magruder will hold their positions in front of the enemy against attack, and make such demonstrations Thursday as to discover his operations. Should opportunity offer, the feint will be converted into a real attack, and should abandonment of his entrenchments by the enemy be discovered, he will be closely pursued.

III. The Third Virginia Cavalry will observe the Charles City Road. The Fifth Virginia, the First North Carolina, and the Hampton Legion (cavalry) will observe the Darbytown, Varina, and Osborne Roads. Should a movement of the enemy down the Chickahominy be discovered, they will close upon his flank and endeavor to arrest his march.

IV. General Stuart, with the First, Fourth and Ninth Virginia Cavalry, the cavalry of Cobb's Legion and the Jeff. Davis Legion, will cross the Chickahominy tomorrow and take position to the left of General Jackson's line of march. The main body will be held in reserve, with scouts well extended to the front and left. General Stuart will keep General Jackson informed of the movements of the enemy on his left and will cooperate with him in his advance. The Tenth Virginia Cavalry will remain on Nine Mile Road.

V. General Ransom's Brigade, of General Holmes command, will be placed in reserve on the Williamsburg Road by General Huger, to whom he will report for orders.

VI. Commanders of divisions will cause their commands to be provided with three days cooked rations. The necessary ambulances and ordnance trains will be ready to accompany the divisions and receive orders from their respective commanders. Officers in charge of all trains will invariably remain with them. Batteries and wagons will keep on the right of the road. The chief engineer, Major Stevens, will assign engineer officers to each division, whose duty it will be to make provisions for overcoming all difficulties to the progress of the troops. The staff departments will give the necessary instructions to facilitate the movements herein directed.

By command of General Lee:
R. H. Chilton
Assistant Adjutant-General[7]

This then was the document that governed Lee's pending campaign against Porter. It was disseminated on the 24th, and we can be sure was carefully studied by the players that night. In our subsequent discussions, we will disregard sections II, III, and V, which do not directly relate to the movements against Porter, and section VI, which contains general instructions applicable to any campaign.

We will now see how each player responded to his instructions and then see how it all finally played out. In general, we may say that the operation started out poorly and then deteriorated. The big day was June 26, 1862.

General Branch was a brigade commander in General A. P. Hill's division. He was to move seven miles up the Chickahominy from A. P. Hill's camp at Meadow Bridge to Half Sink Bridge. When notified by a courier from Jackson that Jackson had begun crossing the Virginia Central Railroad (see map 7), he was to cross the Chickahominy and come down the other side to clear out the Yankees from the approaches to Meadow Bridge so that A. P. Hill, with the rest of the division, could cross. He would rejoin his division for further action. With that, let us turn to the words of General Branch.

> In my written orders it was stated that General Jackson would cross the railroad at 3 o'clock Thursday morning, and allowing one hour for the transmission of the message, I was under arms and prepared to cross at 4 A.M. on Thursday.
>
> Not having received any intelligence from General Jackson, and General Lee's written orders to me being explicit, there was no danger of my making a false movement; but after 8 o'clock in the morning I received from you [Lee] an order in these words: "Wait for Jackson's notification before you move unless I send other orders."

Up to this time my brigade was in the open fields near the banks of the stream in full view of the enemy's pickets on the other side. To deceive them as to my purpose I now marched it back a half a mile in the direction of my camp at Brook church, and marched it into the woods.

At a few minutes after 10 A.M., I received from General Jackson a note, informing me that the head of his column was at the moment of his writing crossing the Central Railroad.[8]

Branch then crossed the river with the enemy pickets retiring before him. When he reached Meadow Bridge, he found that A. P. Hill and the rest of his division had already crossed without waiting for him. Branch then rushed to catch up with the rest of the division.

Now let us turn to the story of A. P. Hill. The plan called for Hill to wait at Meadow Bridge and cross when Branch cleared the opposite side.

My orders were that General Jackson, moving down from Ashland would inform General Branch of his near approach. As soon as Jackson crossed the Central Railroad, Branch was to cross the Chickahominy, and taking the river road, push on and clear the Meadow Bridge. This done I was to cross at Meadow Bridge, and sweeping down to Mechanicsville, open the way for General Longstreet. It was expected that General Jackson would be in the position assigned him by early dawn, and all my preparations were made with the view of moving early. General Branch, however, did not receive intelligence from General Jackson until 10 o'clock, when he immediately crossed and proceeded to carry out his instructions. He was delayed by the enemy's skirmishers and advanced but slowly.

Three o'clock having arrived, and no intelligence from Jackson or Branch, I determined to cross at once...[9]

It was past 3 P.M. when Hill finally crossed Meadow Bridge and proceeded down the river toward Mechanicsville. True to his orders, he began to push the Yankees in Mechanicsville back to their defense position at Beaver Dam Creek and to clear the bridge approaches. By 4 P.M. he was before Beaver Dam Creek, but was not able to clear the bridge approaches until almost 6.

In his post battle report, Hill described his position before Beaver Dam Creek as follows:

The battle now raged furiously along my whole line. The artillery fire from the enemy was terrific. Their position along Beaver Dam Creek was too strong to be carried by a direct assault without heavy loss, and expecting every moment to hear Jackson's guns on my left and the rear of the enemy, I forebore the storming of their lines.[10]

Darkness crept on and by 9 P.M. the firing finally ceased. Only one brigade of the D. H. Hill-Longstreet force succeeded in joining A. P. Hill in the fighting before dark. With that small exception, A. P. Hill had done all the fighting that day.

It was now dark, the end of fighting on June 26, and the guns of Jackson were never heard.

Next we come to General Stuart, Lee's cavalry commander. For the operation to be a success, he had to play a key role. Jackson was tasked with making the key attack of the operation by attacking the Union forces on Beaver Dam Creek on their flank and rear. However, Jackson was new to the area, had just arrived, was unfamiliar with the terrain, and had never reconnoitered the enemy position. Stuart, on the other hand, was intimately familiar with the terrain and had closely reconnoitered the enemy position.

Let us turn to the words of Stuart:

It is proper to remark here that the Commanding General [Lee] had, on the occasion of my late expedition to the Pamunkey, imparted to me his design of bringing Jackson down on the enemy's right flank and rear, and directed that I should examine the country with its practicability for such a move. I therefore had studied the features of the country very thoroughly, and knew exactly how to conform my movements to Jackson's route.... The information obtained then and reported to him [Lee] verbally, convinced the Commanding General that the enemy had no defensive works with reference from that direction ... that his forces were not disposed so as successfully to meet such an attack, and that the natural features of the country were favorable to such a descent. General Jackson was placed in possession of all these facts.[11]

The tasking of Stuart in General Order 75 said nothing about advising, accompanying, or guiding General Jackson. It merely tasked Stuart with taking position on the left of General Jackson's march, and "...General Stuart will keep General Jackson informed of the movements of the enemy on his left and will cooperate with him in his advance."[12]

Stuart's post battle report talks of constant skirmishing with the enemy as he advanced, the reaching and repairing of the bridge over Totopotomoy Creek near Pole Green, and finally, of Jackson going into bivouac for the night at Hundley's Corner. He gives no times.

Now let us turn to General Jackson's activity. The drafter of General Order 75 was mistaken as to Jackson's position on the 25th. He assumed that Jackson was already ensconced at Ashland and directed him to proceed to the Virginia Central Railroad that date. In fact, Jackson's troops only arrived at Ashland throughout the 25th, and hence when he began his march on the 26th, he started from Ashland and not from the Virginia Central Railroad. Thus, instead of having just seven miles to cover to reach his attack point, he had 19 to cover—12 to the railway and seven more to the attack point.

General Whiting's division was at the head of Jackson's column. The march was not uncontested, but as Longstreet had predicted, encumbered by fallen trees across the road, and whatever other obstacles retiring pickets and enemy cavalry detachments could effect. Whiting stated that he reached the last water obstacle that Jackson had to cross before reaching his attack point at 3 P.M. This was Totopotomoy Creek just before Pole Green (see map 8). We will turn to Whiting's words to describe what transpired next:

> At 3 o'clock reached the creek [Totopotomoy], found the bridge in flames, and found a party of the enemy engaged in blocking the road at the opposite side. The Texan skirmishers gallantly crossed and engaged. Reilly's battery, being brought up, with a few rounds dispersed the enemy; the bridge was rebuilt and the troops crossed, continuing on the road to Pole Green Church, and Hundley's Corner. Here we united with Ewell's division, and night coming on bivouacked. A furious cannonade in the direction of Mechanicsville indicated a severe battle.[13]

Whiting did not say what time they went into bivouac. However, Stuart's report indicated that it only took one-half hour to repair the bridge. Thus, the head of Whiting's column was across before 4 P.M., and the distance from the bridge to Hundley's Corner was only about one and one-half miles. Thus, Whiting's column must have reached the bivouac area no later than 5 P.M. To go into bivouac at that time was early indeed as this was June 26, just four days beyond the longest day of the year. We know from other testimony that there was sufficient light for A. P. Hill's battle, which was raging just two to three miles beyond Hundley's Corner, to continue until 9 P.M.

General Trimble was a brigade commander in Jackson's corps and was at Hundley's Corner the afternoon of the 26th. He subsequently stated:

> At 4 P.M. heard distinctly the volleys of artillery and musketry in the engagement of General Hill with the enemy. Before sundown the firing was not more than two miles distant, and in my opinion we should have marched to the support of General Hill that evening.[14]

Trimble's comment was not made years after the event when his memory could have been faulty, but was contained in his official report of the battle submitted on July 18, 1862, just 32 days after the event.

If one could describe the Confederate implementation of General Order 75 on June 26, 1862, in a single word, that word would be "fiasco." The intent was to hit Porter's 30,000 in front with 30,000 while Jackson hit them in the flank and rear with another 30,000. Only 10,000 of the Confederates ever got into action; this was the division of A. P. Hill augmented by the brigade of General Ripley. The Confederate casualties for the day were 1,484, while the Union's were 361.[15]

If there was hope of a Napoleonic coup for the Confederates on the morning of June 26, by nightfall that hope was gone with the wind. McClellan, who was with Porter at the time, recognized the threat being posed by Jackson on his flank, and ordered a withdrawal of the Union forces during the night.

The new position Porter occupied was just a couple of miles further down the Chickahominy and was centered on a place called Gaines Mill. The position was semi-circular in shape and thus less amenable to being flanked. Of even greater importance, it covered four bridges on the Chickahominy. These were Duane's, Woodbury's, Alexanders's, and Grapevine. These four bridges had Union controlled territory on the far side. Thus, Porter could be reinforced if desired or retreat if required.

The battle was continued on the morrow, June 27, 1862, in the so-called battle of Gaines Mill. This battle offered no prospects of a Confederate "masterful stroke" like that attempted on the previous day. It was a slugfest in which superior numbers must ultimately prevail. Porter was first reinforced during the day via the bridges in his back, and then retreated over them during the night. Final casualties for the day were Confederates 8,751 and Union 6,837.[16] The Confederates were thus given another bloody nose as Porter escaped intact.

The Confederates did, however, win a strategic victory. The Union army south of the Chickahominy was now cut off from its supply base at White House, and White House itself was doomed. If McClellan previously had wanted to transfer his supply base to the James, now he must. The initiative had passed to Lee. The siege of Richmond was lifted, at least for a time.

Now we come to the key question in our chapter. Could Lee have achieved a smashing victory at Mechanicsville that would have obviated the requirement for the Battle of Gaines Mill had General Order 75 been written differently?

Major General Fitz John Porter USV: A worthy opponent in the Seven Days Battles.

First, we must note that General Order 75 was complicated and confusing in the extreme. It never actually ordered Jackson to attack anyone, much less specify that he was to attack the enemy position at Beaver Dam Creek in the flank and rear on June 26. Thus, by not making such an attack, Jackson did not technically violate his orders.

Let us now redraft the order based on the principles enumerated in the introduction. In doing so, we must keep the basic plan intact. In our redraft, we will (1) simplify the order, (2) specifically indentify the required actions and desired objectives, (3) widen the discretion of the key players, (4) reduce the number of decision makers and their reporting to the commanding general, and (5) eliminate most contingent action situations.

Our new order is as follows:

General Orders
No. 75
HDQTRS Army of Northern Virginia
June 2, 1862

I. My intention is to destroy the Union forces north of the Chickahominy on June 26, 1862.
II. I propose doing the above by having the forces of Longstreet, A. P. Hill and D. H. Hill crossing the upper Chickahominy, seizing Mechanicsville, and confronting the Union forces at their Beaver Dam Creek defense line, while the forces of General Jackson attack them from behind and in their right flank.

Assignments are as follows:

General Jackson: Employing all the time between the receipt of this order and the attack, as he deems appropriate, will approach the Union line at Beaver Dam Creek via Pole Green, and will attack it in flank and rear. Such attack is to take place on Thursday, June 26, after 8 A.M. but otherwise as early in the day as feasible.

General Stuart: Will report to General Jackson with all of the cavalry not assigned elsewhere in this order, not later than the morning of June 25 and will remain under his command until further notice.

General A. P. Hill: Will cross the Chickahominy at daylight Thursday, June 26 via Half Sink or Meadow Bridge or both, as he deems appropriate. He will then proceed down the Chickahominy and (A) clear the approaches to the Mechanicsville bridges so that the forces of Longstreet and D. H. Hill can cross, and (B) seize Mechanicsville.

General Longstreet: You will command the forces of your division and that of D. H. Hill to cross the Chickahominy on June 26 when you deem appropriate. You will then take command of the three divisions; i.e., yours, D. H. Hill's and A. P. Hill's, and complete the seizure of Mechanicsville. You will then demonstrate before the Union line on Beaver Dam Creek, but not attack until ordered so by me. The deployment of your troops will be such that you will be able to transition into a full scale attack at any time after 8 A. M. when so ordered.

Generals Jackson, Longstreet, A. P. Hill, and D. H. Hill are authorized to deviate from these instructions and use your own judgment under either of two circumstances: (A) You encounter an unforeseen opportunity, the exploitation of which offers greater gain; (B) You encounter a situation wherein you consider to proceed would be a waste of the lives of your men without any realistic prospect of success.

My headquarters will be at the Mechanicsville Bridge until further notice.
By command of General Lee:
R. H. Chilton
Asst. Adjutant General

Had this been the wording of General Order 75, would it all have turned out differently?

4

Malvern Hill

In chapter 3, we saw how Lee seized the initiative in the battle of Mechanicsville. This was the beginning of the famous "Seven Days Battles," which ultimately transferred the scene of action from the gates of Richmond to the doorstep of Washington.

In chapter 3, we covered the first three days of the seven days culminating in the battle of Gaines Mill. In this chapter, we will pick up and cover the remaining four days, ending in the battle of Malvern Hill.

One might conclude that the Seven Days Battles constituted a masterfully conceived and brilliantly executed campaign. To the contrary, it was a series of blunders and missed opportunities leavened by a very large degree of incompetence.

In order for us to understand the final Confederate fiasco at Malvern Hill, we will now go back and start with day four, Saturday, June 28, 1862.

Saturday, June 28, 1862

Saturday the 28th was a day of assessment for Lee. Having won at Gaines Mill, he had complete control of the area above the Chickahominy all the way to Long Bridge. McClellan's supply line had been cut. The Confederates not only seized the segment of the railroad between the Chickahominy and White House, but White House itself. McClellan's troops had to eat; the horses had to have fodder; and he needed a continuous replenishment of ammunition, medical supplies, and all the other consumables an army requires. The York Peninsula on which McClellan was located was deep in enemy territory. All his supplies had to come by sea. He needed a supply base on a navigable waterway that extended all the way to Washington, and he needed it fast.

Lee considered that McClellan had four options (see map 9). The first option was to move down the south side of the Chickahominy to a point beyond Confederate control, cross over, and attempt to recover White House and the rail line.

The second option was for McClellan to retreat back down the Yorktown Peninsula from whence he came and establish Yorktown as his main supply base. To do this, however, would be admitting defeat. Furthermore, it would eliminate all possibility of resuming the siege of Richmond in the near future. Yorktown was over 60 road miles from Richmond. Thus, the supply for any siege forces, once unloaded at Yorktown, would have to travel 60 miles by horse-pulled wagons over roads that were at best bad and sometimes impassable.

**Map 9
Malvern Hill — McClellan's Options**

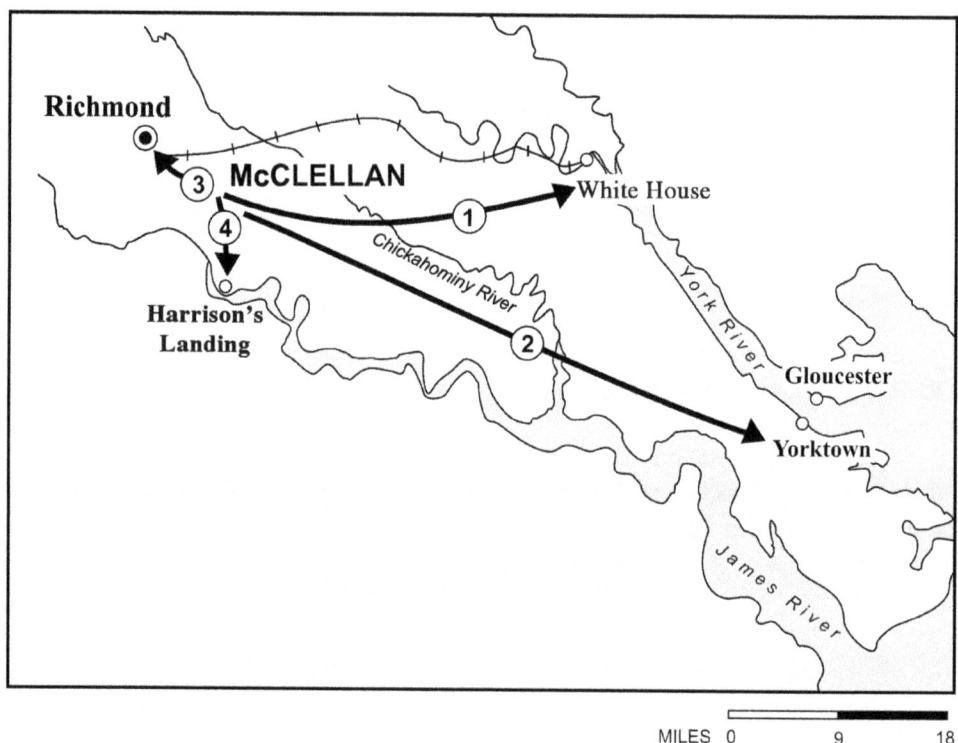

The third option available to McClellan constituted a nasty possibility for the Confederates. Once the Union forces had withdrawn to the south side of the Chickahominy after Gaines Mill, the entire Union army was on the south side, and only one-third of the Confederate army was. This third was in the earthworks guarding Richmond. The Union army thus outnumbered the defenders by almost four to one. If McClellan decided to make an all out attack on Richmond and worry about his supply base later, success was almost assured. However, this decision would have to be made by a daring commander, and Lee, who knew McClellan well, knew that this he was not.

The fourth option available to McClellan was to establish a supply base on the James River as close as possible to Richmond. This would be Harrison's Landing, which was just 14 miles from Grapevine Bridge on the Chickahominy, and just 20 miles from Richmond. An added benefit of Harrison's Landing was that the army here could be supported by the powerful flotilla of gunboats on the James River. In fact, any Confederate force even approaching Harrison's Landing would be subject to naval bombardment even before it closed with the Union land forces.

McClellan chose option four. He planned to withdraw the entire army to Harrison's Landing where it would be rested, refurbished, and massively reinforced; and then he would resume the siege and capture Richmond.

By the end of the 28th, Lee had correctly divined McClellan's intentions and was drawing up his plan of action.

McClellan's Retreat Route

Before we proceed to the actions of Sunday, June 29, 1862, let us take a look at the route McClellan was to follow to reach Harrison's Landing, and at some of the problems the retreat entailed. First of all, the retreat was not just a matter of choice. Once Lee achieved mastery north of the Chickahominy, the Union siege forces were flanked. Lee could cross by downriver bridges and place his army behind the siege forces. McClellan had to pull his siege forces back.

The retreat was no small matter. McClellan had over 100,000 men, about 5,000 horse-drawn wagons, over 300 pieces of field artillery, numerous heavy siege guns, and mountains of supplies and equipage. In addition, he had thousands of sick and wounded in field hospitals. Most of the army was in the siege force before Richmond facing the Confederate earthworks. The remainder was just south of the Chickahominy, having crossed the river after the Union defeat at Gaines Mill. Moving this massive force across the peninsula while being pursued by a resourceful and determined enemy was no small undertaking.

First, the troops in the siege positions had to be withdrawn eastward before proceeding south. The north-south retreat route extended approximately 14 miles from Grapevine Bridge to Harrison's Landing (see map 10). Not all of the troops, of course, used the entire route, as those withdrawing from their siege positions entered the route at various intermediate points.

Shortly after leaving the Grapevine Bridge and proceeding south, the route quickly ran into a morass called White Oak Swamp. This in turn was bisected by a small stream that ran west to east. There were a number of paths and primitive roads in the swamp, but only one north-south road that had a bridge over the stream. One might find his way through the swamp on foot or horseback, but all vehicular traffic had to take the single road with the bridge. The bridge was known as White Oak Bridge.

One exited the swamp near the mid point of the route and came to a small settlement called Glendale. Glendale was an important road junction. Two roughly east-west roads joined the north-south retreat route at this point. From Glendale south to Harrison's Landing there were no additional swamps or bridges. However, as one continued south from Glendale for about two miles, one came to a hill that rose to an elevation of about 100 feet. This was Malvern Hill. Malvern Hill was on the James River. Upon passing over Malvern Hill, the road turned and followed the James downstream for about three miles, where it reached Harrison's Landing.

Any experienced military man could see the defensive possibilities of Malvern Hill. It could not be attacked on the south side because of the James. The crest of the hill was sufficiently wide to mount large numbers of cannon, and the area in front was wide open and provided no cover for an attacking force. The artillery had an unobstructed field of fire. Confederate General D. H. Hill advised Lee, "If General McClellan is there, we had better leave him alone."[1]

In summary, there were three militarily significant points on the 14 mile north-south retreat route to Harrison's Landing. The first was the bridge over the stream in the center of White Oak Swamp. The second was the road junction at Glendale where large numbers of troops could be quickly congregated. Last was Malvern Hill, which could be easily converted into a nearly impregnable defense position.

Map 10
Malvern Hill — Situation at Midnight, June 28, 1862

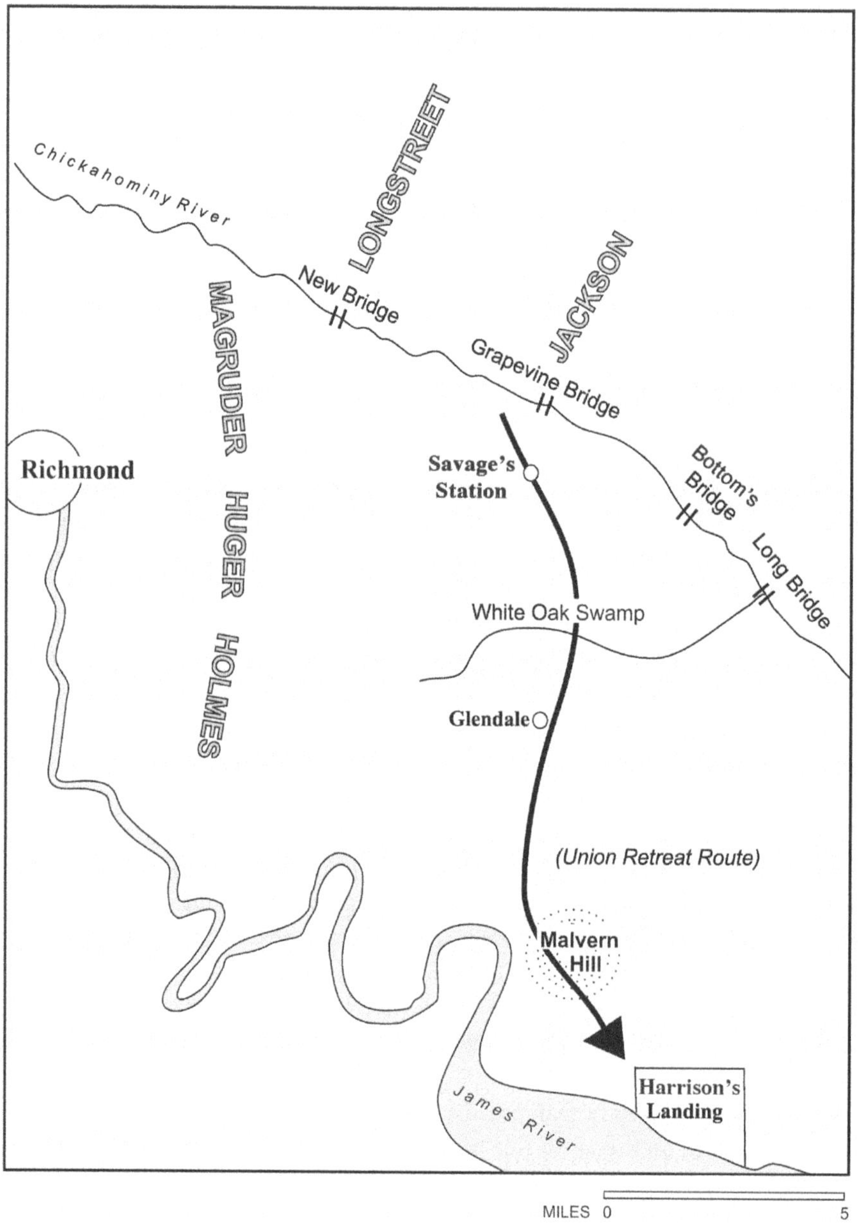

The Two Main Players

Before continuing on to Sunday, June 29, 1862, let us look at two of the main players in the subsequent events: Major General John B. Magruder and Colonel Robert H. Chilton.

General Magruder was one of those "bigger than life" individuals one occasionally encounters. He was handsome, a natty dresser when in mufti, and gregarious. His avocation was the theater, and he spent his spare time writing songs and staging amateur theatrical productions. He was known to his friends as "Prince John."

Magruder was not one of the "boy wonder" generals of the Civil War. In fact, at the time of the secession of his native Virginia, he was 54 years old and had been an officer for 31 years. He had graduated from West Point way back in 1830, just one year after General Lee and two after President Jefferson Davis. Magruder served in the Seminole War, acquitted himself admirably in the Mexican War, and fought Indians on the frontier. He chose, however, to cast his lot with the South. At the outset of the war, he resigned his commission and accepted a colonelcy in the newly forming Confederate Army.

In one of his earlier assignments, he was put in charge of the so-called "District of Yorktown" on the York Peninsula. In this early phase of the war, the Union clung to toeholds in southern Virginia. These

Major General J. B. Magruder CSA: "Prince John" to his friends.

included Fort Monroe at the tip of the York Peninsula and the nearby naval facility at Newport News. Both were near Magruder's command at Yorktown. Magruder's small command inevitably came in contact with the nearby Yankees in one of the first battles of the war. This was the battle of Big Bethel, which took place on June 10, 1861. Although the casualties were laughably small when compared with what was to come, the battle seemed enormously important at the time and achieved wide publicity. The casualties were: Union—eighteen dead, 53 wounded, five missing; Confederates—exactly one dead and seven wounded.

Big Bethel was a resounding Confederate victory and the man in charge was Colonel John Magruder. He was a hero throughout the South and was quickly promoted to brigadier general and then major general. Magruder was now one of the senior officers in the newly formed Confederate Army.

Real war came to the York Peninsula in the spring of 1862 when McClellan transported the Army of the Potomac by sea to Fort Monroe to attack Richmond from the east. Magruder, now a major general, was in command of all the small, scattered Confederate forces on the peninsula. His task was now to hold McClellan's huge army off long enough for General Johnston to bring the main Confederate army south for the defense of Richmond.

Magruder's task seemed impossible. McClellan's army of over 100,000 should have been easily able to brush Magruder's tiny force aside and march on to Richmond. Here, Magruder's theatrical inclinations came to his aid. By a series of ruses, he managed to convince McClellan that his forces defending Yorktown were vastly greater than they actually were. This caused McClellan to shift from mobile warfare to a ponderous and time-consuming siege mode. This gave Johnston time to gather his forces, and McClellan never entered Richmond. If Magruder was a hero before, he was *the* hero now. His stock was at its peak. Now we come to the Seven Days Battles culminating in Malvern Hill.

Magruder's successes to date could be largely attributed to the fact that he was at the right place at the right time. In the Seven Days Battles, he had always been at the right place at the wrong time, or vice versa. He did not live up to Lee's expectations and had to go. He would play no further role in the eastern theater for the remainder of the war.

Colonel Robert Chilton was not a wartime product but, at the outset of the war, a seasoned professional officer from the old army. Chilton had graduated from West Point in 1837 and thus, in 1861, had 24 years of experience as an officer. He had fought Indians and served with great distinction in the Mexican Wars. Among his accomplishments, during the battle of Buena Vista, while under severe fire, he picked up the wounded Colonel Jefferson Davis and carried him to safety.

At the outset of the Civil War, Chilton resigned his commission as major in the regular army and accepted a commission of colonel in the Confederate Army. The fact that the man whose life Chilton saved was about to become president of the Confederate States did not hurt Chilton's prospects.

On June 4, 1862, Chilton was assigned as Lee's chief-of-staff. He remained in this position until April 1, 1864. Consequently, Chilton's tenure covered a vital segment of the war that included the Seven Days Battles, second Bull Run, Antietam, Fredericksburg, Chancellorsville, and Gettysburg.

Chilton, who was 47 years old when assuming office, was closer in age to Lee than the other staff officers and got on well with Lee. In fact, some looked upon Chilton not only as a subordinate of Lee but as a friend of Lee. Chilton named his third child Robert Lee Chilton.

Lee appeared to have great confidence in Chilton and, at times, when more than a simple order was required to a distant subordinate, Lee would send Chilton to speak for him. The subordinate generals readily acquiesced in accepting Chilton's word for Lee's. This could be explained in part by the facts that, unlike the youthful Sandy Pendleton, whom we met in chapter 2, Chilton was probably older than the general he was talking to, probably had graduated from West Point earlier, and was probably senior to him in the old army.

Written orders emanating from Lee's headquarters were usually signed by Chilton. We do not know the mechanics by which a verbal order by Lee was transformed into a written order by Chilton. Did Lee dictate? Did he merely describe what he wanted and leave the words to Chilton? Did he read and approve Chilton's order before it went out? We do not know. It could have been a combination of all three.

In any event, the written orders sent out over the signature of Chilton were not of the quality typical of Grant. Chilton was no master order writer. His orders sometimes lacked something as elemental as a time of origin. Orders signed by Chilton managed to appear in three of the 14 cases in this book, including Malvern Hill.

Sunday, June 29, 1862

When the Union forces retreated across the Chickahominy after Gaines Mill, they destroyed the bridges behind them. Thus, as of the 27th, all of the bridges downriver from Mechanicsville were down. The Chickahominy was not overflowing its banks into swampland at this time and the restoration of the bridges was a fairly simple matter. The river was only 40 feet wide, and most bridges could be reconstructed within 24 hours once tools and work crews were available. As of the morning of the 29th, New Bridge had already been restored and work was in progress on Grapevine Bridge.

Lee's situation at midnight on the 28th was as follows (see map 10): Longstreet / A. P. Hill (23,000) were camped on the north side of New Bridge; Jackson / D. H. Hill (28,500) were camped on the north side of the Grapevine Bridge. The three commands in the earthworks defending Richmond south of the Chickahominy were: Magruder (13,000) northernmost, Huger (9,000) center, and Holmes (6,500) southernmost on the James.

As for the Union this morning, the Union army was strung out on the north-south route all the way from Savage's Station (just south of Grapevine Bridge) to Harrison's Landing.

Lee's plan was to hit the retreating column at Glendale, separate the column north of Glendale from that of the one south, and destroy the northern segment. To this end he gave the following orders. Longstreet's force was to cross New Bridge, proceed south, and arrive at Glendale on the 30th. Jackson's force was to cross Grapevine Bridge and press the rear guard of the retreating Union force from the north toward Glendale. Huger was to take a road that met Longstreet's force at Glendale. Magruder and Holmes were to vacate their entrenched positions and proceed east to attack the retreating Union column in the flank.

All Confederate columns moved out on the morning of the 29th except Jackson's, which had not yet succeeded in restoring the Grapevine Bridge.

The first Confederate column to contact the enemy was that of Magruder. Magruder entered into low level combat soon after stepping off, but did not meet the main Union force, which constituted the retreat rear guard, until late in the afternoon. The two forces met at a place called Savage's Station, a station on the Richmond-York Railroad. Savage's Station was just two miles from Jackson's force at Grapevine Bridge. However, the restoration of Grapevine Bridge was still not complete, and Jackson's force did not get over the river until early on the 30th when Magruder's fight was long over. This constituted the second instance in the Seven Days Battles wherein Jackson's large force sat idle when within the sound of a battle.

Magruder's fight with the Union rear guard is known as the battle of Savage's Station. The fight was inconclusive and the Union column continued on south into White Oak Swamp. The casualties for the battle were: Union —1,038; Confederate — 473.

The Confederates did acquire one dubious advantage as a result of the battle. The retreating Union column abandoned a field hospital containing 2,500 Union sick and wounded to the Confederates.

Monday, June 30, 1862

This was to be the day on which Lee would complete the execution of his master plan. He would cut the retreating Union column at Glendale and destroy the component to the north.

By early morning of the 30th, the rear guard of the retreating Union column had already cleared White Oak Bridge and the head of the column had already reached Harrison's Landing. By about 11 A.M., the Longstreet / A. P Hill column of some 23,000 was already in position to assault the Union column at Glendale.

Huger's column of 9,000 was approaching Glendale on a road converging from the northwest. Jackson's command, including that of D. H. Hill and containing 28,500, was approaching from the north through White Oak Swamp and had arrived at White Oak Bridge, which the retreating Union column had destroyed. At this point, the three converging Confederate columns were all within three miles of each other (see map 11). Combined, they totaled 60,500, vastly outnumbering the Union forces north of Glendale. At this point, it appeared that Lee's plan was a smashing success. They could hardly fail to

Map 11
Malvern Hill — Battle of Glendale, June 30, 1862

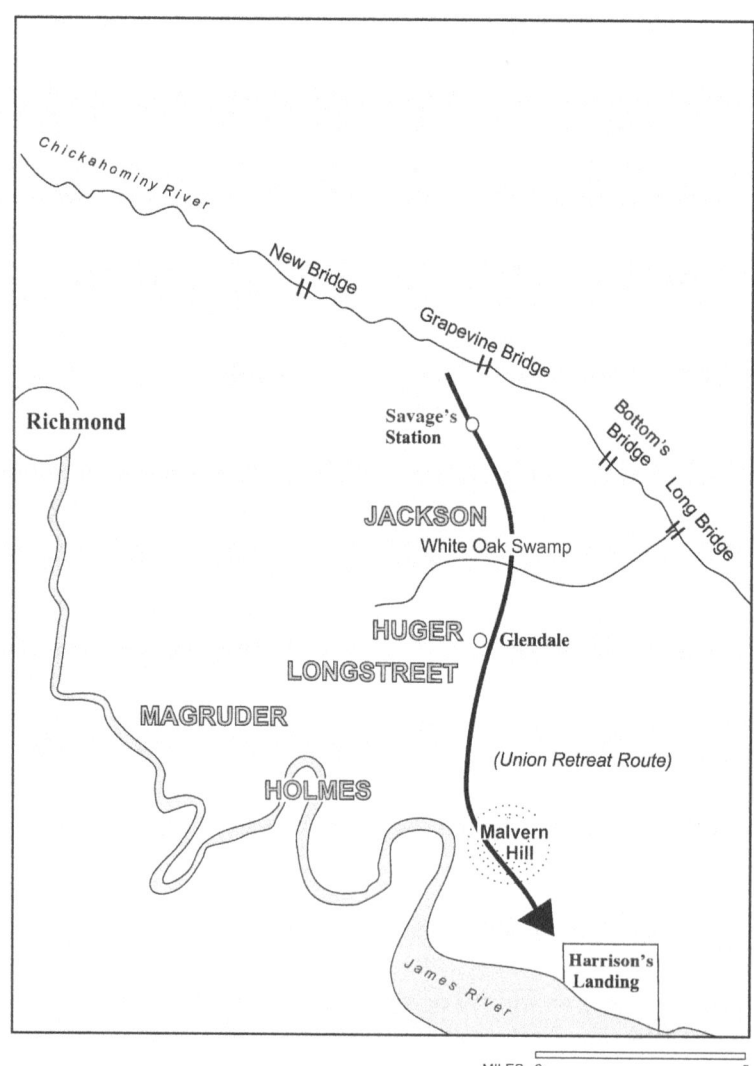

destroy the opposing Union force. Fail, however, they did, and a near certain victory slipped through Lee's fingers.

When Longstreet's force arrived at Glendale, he waited for signs that Huger's and Jackson's forces were close enough to cooperate before he attacked. At 2:30 P.M. he heard cannon fire. He assumed it was from Huger or Jackson and opened the battle with an attack. However, Longstreet was destined to be disappointed. The converging columns of Jackson and Huger were never to arrive in time to participate in the battle.

When Jackson arrived at White Oak Bridge, he found the bridge destroyed and Union forces strongly posted on the opposite bank. He sent his cavalry up and down the stream to look for a suitable crossing point. They quickly returned and reported that there was a nearby ford completely suitable for infantry crossing.[2]

For whatever reason, Jackson elected not to cross and, thus, his large force played no part in the battle of Glendale. Jackson's skirmish at the bridge is sometimes called the battle of White Oak Swamp, but the casualties were insignificant. This constituted the third time in the Seven Days Battles that Jackson's large force sat out a battle when within the sound of guns. Jackson was subsequently criticized for his actions at Glendale by General D. H. Hill, his friend and brother-in-law,[3] and by General Longstreet.[4]

Huger's lack of participation in the battle of Glendale was a little more justified than Jackson's. He simply moved at a snail's pace because his movements were retarded by the enemy felling trees across his route.

While Longstreet was fighting the battle of Glendale, Holmes's force, which was south of Longstreet and moving along the James River, came in sight of Malvern Hill. By this time, it came under bombardment by Union gunboats. This the troops found particularly demoralizing. The great majority of field cannon of the time fired shells of 12 pounds or less. The gunboats fired 100-pound shells. The recipient troops referred to these incoming shells as "lamp posts."[5]

Holmes took one look at Malvern Hill and decided that it could not be taken with his force. Actually, he was probably wrong as at this time the hill was only lightly garrisoned. In any event, Holmes sent to Longstreet for reinforcements. Longstreet in turn ordered Magruder, who was heading for Glendale but still some distance away, to go to Holmes's assistance. Before Magruder even succeeded in connecting with Holmes, he received an order from Colonel Chilton, Lee's chief-of-staff, to go and support Longstreet. Before Magruder could reach Longstreet, darkness ended the day's activity.

Today was the day the Confederates expected to win a smashing victory. They could have and should have but did not. Had Longstreet cut the Federal retreat off at Glendale or had Holmes seized Malvern Hill, they would have succeeded. Longstreet failed and Holmes never tried. The commands of Huger and Holmes never got into action, the command of Jackson barely so, and the command of Magruder wasted its time marching and counter-marching but never reaching either destination. The casualties for the day were: Union — 3,707; Confederate — 3,673.

As darkness fell, the Union retreat continued with the rear guard passing through Glendale en route to Malvern Hill and thence to Harrison's Landing.

Before we leave the 30th, let us look at how McClellan saw the day. At 7 P.M. he sent the following dispatch to the War Department:

Hon. E. M. Stanton

Another day of desperate fighting. We are hard pressed by superior numbers. I fear I shall be forced to abandon my materiel to save my men under cover of the gunboats. You must send us very large re-enforcements by way of Fort Monroe, and they must come very promptly. My army has behaved superbly, and have done all that men could do. If none of us escape, we shall have at least done honor to the country. I shall do my best to save the army. Send more gunboats!

Geo B. McClellan

Major General Commanding[6]

Tuesday, July 1, 1862

Now we come to the day of our subject, the battle of Malvern Hill. As the sun rose on Tuesday, July 1, 1862, Lee's chances of a one-sided victory were gone with the wind. It was no longer possible to cut off a piece of the retreating Union army. McClellan's 5,000-some wagons were now safely parked at Harrison's Landing, his army united, and his infantry and artillery busy establishing themselves on Malvern Hill to receive any further effort of Lee. In addition to McClellan's army, three powerful gunboats were standing by on the James to render support.

Lee could have and should have called it quits at this point. He had lifted the Union siege of Richmond and he had seized the initiative. He could now march north toward Washington and thus force the Union to withdraw McClellan's army from the York Peninsula to come to the defense of Washington. This is, of course, what ultimately happened — but not yet.

Quitting now was not in Lee's nature. He had set out to destroy the Union army, and but for some bad coordination since the seven days began, should have done so. He had come so close that he could taste it. Now that, realistically, the opportunities were gone, he would make one more mighty effort and this time it would be coordinated properly.

Lee could see that to make a frontal infantry attack across the open ground before Malvern Hill and then up the hill was suicidal. Longstreet, however, convinced him that a crossfire of artillery against the hill could soften up the position for the infantry.

The plan was to first bring up the artillery for the crossfire, and then follow up the barrage with an advance of infantry. The infantry was to be lined up as follows: Jackson — Magruder — Huger — (gap) — Holmes. The Longstreet / A. P. Hill force, which had done all the fighting on the previous day, was to be held in reserve.

The preparations for the attack began, and it all started badly. Magruder was not yet on the scene, so Lee sent him an order to march down the Quaker Road and form on Jackson's right. Maps of the time and area were notoriously bad and there were few road signs. So, to prevent any possibility of a slip up, Lee sent Magruder a guide who had lived in the area his entire life.

After proceeding a short distance with Magruder, the guide pointed to an overgrown, seldom used country road and said that this was the Quaker Road. This did not seem right to Magruder, so he inquired of other local inhabitants. All affirmed that this was the Quaker Road and that there was no other. So, Magruder dutifully marched down what proved to

4. Malvern Hill

be the wrong road. As it turned out, the road that Lee thought was the Quaker Road was known to the locals as the Willes Road. By the time Magruder learned of his error and retraced his route, Lee's divisions were already lining up for the attack, and Huger had formed on Jackson's right. Magruder did the only thing possible and formed on Huger's right. The order for the attack was finally disseminated to the division commanders late in the afternoon. It was as follows:

> General (Various):
> Batteries have been established to act upon the enemy's line. If it is broken as is probable, Armistead, who can witness the effect of the fire, has been ordered to charge with a yell. Do the same.
> R. H. Chilton
> A. A. G.[7]

Armistead was a brigade commander in Huger's division and was near the center of the line, and thus in the best position to observe the effects of the preparatory artillery barrage. Nevertheless, it is strange that Lee would delegate to a junior subordinate the crucial decision of when to make the attack. One would assume that Lee himself would occupy the best observation position and make this vital decision.

The order was ambiguous and strange in two further aspects. It read "If it is broken [the line] ... Armistead ... has been ordered to charge...." Did this mean that if the line was not broken by the artillery barrage there would be no charge? Also strange was the proviso that the other units would attack when they heard the shout of Armistead's men. A battlefield is a noisy, confusing, and smoky place, and it was more than a half a mile from one end of the attack line to the other.

Lee was now ready for the preparatory artillery barrage. As each Confederate battery came out of the woods to take position, it was overwhelmed by a shower of shells from Malvern Hill. It was quickly apparent that the bombardment was not going to succeed and was not going to "break" the Federal line.

General D. H. Hill noted the ineffectiveness of the bombardment and later said, "I wrote to General Jackson that the condition upon which the order was predicated was not fulfilled and that I wanted instructions. He replied to advance when I heard the shouting."[8]

Shortly afterward, Hill erroneously thought he heard the shout of Armistead's men and ordered his division forward. The brigades to his left and right did not move. Hill advanced alone. Meanwhile, Magruder, whose troops were still arriving, was delivered a copy of the order signed by Colonel Chilton and ordered his brigades into action piecemeal as they arrived.

In short, the attack was a fiasco. The artillery preparation was useless. The attack was not coordinated and some units never stepped off at all. The attack entailed numerous casualties and never even came close to success. It was an unnecessary and unproductive bloodbath. The casualties were: Union — 3,000 (estimated); Confederate — 5,355.

But yet, D. H. Hill was later to write: "If one division could effect this much, what might have been done had the other nine cooperated with it."[9]

In summary, had the Confederate attack been coordinated, it just might have succeeded. It was not coordinated because of faulty order writing. The troops did not fail, the order writers did. And lastly, the Federals abandoned Malvern Hill the very night of the battle. Had the Confederates waited one day, they could have had it for nothing.

The Aftermath

The battle of Malvern Hill completed the so-called Seven Days Battles. McClellan's encampment at Harrison's Landing was in a near invulnerable position, but the location was unhealthy. His casualties due to sickness began to mount alarmingly.

General Magruder was to have no further role in the eastern theater. He was initially transferred to the newly created Department of the Trans Mississippi, but shortly thereafter was recalled and assigned to command the District of Texas. Here he remained until the end of the war.

Shortly before Magruder's arrival in Texas, the Union seized the Port of Galveston. Galveston had been exceedingly important to the Confederacy as a terminus for blockade runners, and Magruder resolved to have it back. Magruder's theatrics and proclivity for the imaginative again came to his aid. Galveston was located on an island connected to the mainland by a causeway. Magruder planned a land-sea attack. He would attack both the port and the Union warships anchored therein.

Magruder did not have ironclad warships nor the means to make them. He would use what he had — bales of cotton. Magruder planned to duplicate Washington's famous feat of crossing the Delaware on Christmas Eve and assaulting the drunken, sleeping Hessians at dawn on Christmas Day.

Magruder planned his surprise attack for dawn on New Years Day, 1863. The troops and the "cottonclads" stealthily approached during the hours of darkness. As soon as all was ready, Magruder himself fired the first shot saying, "Now boys, I have done my part as a private, and will now go and attend to that of general."[10]

The attack was a resounding success. Not only did Magruder capture Galveston, but he did what no other Confederate had done to date. He captured a major U.S. warship, the *Harriet Lane*.

Casualties for the Confederates were 12 killed and 70 wounded. The casualties for the Union, including prisoners, were at least ten times as great. Magruder had achieved a resounding victory. He received the following congratulatory message from President Davis:

> Richmond, VA Jan 28, 1863
> Major General J. Bankhead Magruder, Galveston, Texas:
> My Dear Sir:
>   I am much gratified at the receipt of your letter of January 6th, conveying to me the details of your brilliant exploit in the capture of Galveston and the vessels in the harbour. The boldness of the conception and daring and skill of its execution crowned by results substantial as well as splendid. Your success has been a heavy blow to the enemy's hopes, and I trust will be vigorously and effectively followed up.
>   It is hoped that your prudence and tact will be as successful as your military ability etc. ...
>   Jefferson Davis[11]

Prince John was back!

The District of Texas saw little action and received little publicity for the remainder of the war. At war's end, Magruder fled to Mexico. He offered his services to Emperor Maximilian, the puppet emperor who was put up by the French and who was then trying to put down a revolt.

When Maximilian fell, Magruder returned to his now adopted state of Texas and settled in Houston. There he died in 1871 at the age of 63.

After the battle of Malvern Hill, Colonel Chilton was destined to serve as Lee's chief-of-staff for 20 more months. During the 20 months, Chilton was involved in two transactions that could have shattered Lee's faith in him.

The first of these transactions was the so-called "Lost Order." During Lee's first invasion of Maryland in late summer 1862, Lee decided to divide his army to capture the Union garrison at Harpers Ferry in his rear before he continued north. He anticipated that he could conclude the operation and reunite the army before the enemy realized it was divided. Chilton drew up the necessary orders and couriered a copy to each participant. The copy sent to General D. H. Hill was somehow lost and recovered by the Union. General McClellan, the Union commander, now had the means of destroying the Confederate divided army and came within minutes of doing so at the battle of Antietam.

No one at the time and few since pointed the finger at Chilton as being responsible for the loss. It was generally conceded that the loss occurred after the delivery of the message and Chilton was only responsible for the drafting and delivery. But was Chilton really innocent? One can easily visualize actions on his part that would make him culpable. For a message as sensitive as this, the courier should have been instructed to hand the message personally only to General Hill or his AAG, Colonel Ratchford. Did Chilton so instruct the courier? If not, the courier could have handed it to some lowly member of Hill's staff who subsequently lost it. A message of this sensitivity should have been receipted. If a receipt were required and not received, Chilton should have known within hours that the message had not been delivered. Did Chilton fail to require a receipt? In any event, Lee never blamed Chilton and continued to have the highest confidence in him.

The second transaction that could have shattered Lee's confidence in Chilton occurred on May 2, 1863, during the battle of Chancellorsville. In this instance, Lee was commanding the main Confederate army at Chancellorsville and General Early was commanding a smaller Confederate force five miles away at Marye's Heights. Lee sent Chilton to General Early with the message that Early was to remain where he was or join him upon his discretion. Chilton convinced Early that Lee's order to join him was peremptory. Early, against his better judgment, abandoned his position. However, by a series of lucky circumstances, the order was corrected and Early was able to rejoin his position without damage to the Confederate cause. Factors that may have caused Lee to overlook Chilton's actions in this instance were: (1) No harm had been done; (2) Chancellorsville ended in Lee's greatest victory.

Chilton's promotion to brigadier general was finally approved by the Confederate senate on February 16, 1864. With promotion, Chilton was assigned to the position of inspector general of the Army of Northern Virginia. Chilton's new position required that he work in the War Department in Richmond and so, in February 1864, Lee and Chilton parted. Chilton held his new position until the close of the war.

Following the cessation of hostilities and the disbandment of the Confederate Army, Chilton moved to Columbus, Georgia, where he became president of a local manufacturing company. Chilton died in Columbus on February 18, 1879, at the age of 63.

5

The Second Battle of Bull Run

Our next case where the outcome of a battle was influenced by faulty battle order writing was the second battle of Bull Run, or as it is known in the South, the second battle of Manassas.

The battle of Mechanicsville on June 26, 1862, was the first day of the so-called "Seven Days Battles" that completely transformed the situation on the York Peninsula. At the outset of the battle, McClellan's Peninsular Campaign had brought him to the very gates of Richmond and it seemed all but certain that the fall of Richmond was imminent. In the words of Confederate General D. H. Hill, the Seven Days Battles "resulted in lifting the young Napoleon [McClellan] from his entrenchments around that city and setting him down on the banks of the James, 25 miles farther off, with a loss of 51 pieces of artillery, 27,000 stands of arms, and 10,000 prisoners."[1] At the conclusion of the Seven Days Battles, the initiative had passed from the Union army of General McClellan to the Confederate army of General Robert E. Lee.

In Washington, the Peninsular Campaign was now deemed a failure and a new strategy was in order. On June 26, 1862, just as the Seven Days Battles were about to begin, Washington ordered the formation of a new Union army in northern Virginia. This army was to consist of the two corps that had been fighting Jackson in the valley before Jackson joined Lee, and the corps of McDowell. The new army was to be called the Army of Virginia, and it was to be commanded by General John Pope. Its initial task was to facilitate McClellan's capture of Richmond by attacking Richmond from the north and west. However, with McClellan's defeat in the Seven Days Battles and the passing of the initiative to Lee, all this had to be changed. A new strategy was in order.

The new strategy called for withdrawing McClellan's Army of the Potomac by sea to the Washington area. It would then unite with the Army of Virginia, and the new super army would then hopefully steamroll over Lee's army and take Richmond by a land campaign from Washington to Richmond. The new task of the Army of Virginia became defensive. It had to protect Washington until the two armies united.

There was one big vulnerability in the new strategy. Withdrawing McClellan's army by sea was no small task. In fact, it was an amphibious operation of a magnitude that was not to be surpassed until the D-Day invasion of World War II. McClellan had over 120,000 men; 30,000 horses; almost 5,000 wagons; 44 artillery batteries; thousands of sick and wounded; and mountains of ammunition, commissary supplies, forage, and medical supplies. There was the danger that Lee would attack and overwhelm the Army of Virginia while McClellan's army was in the process of being evacuated.

A fact that enhanced the danger was that the Army of the Potomac could not be picked up and landed in fighting form. Some ships were configured to carry personnel, others horses, others cannon, others bulk cargo. Thus, the army had to be picked up by type and reassembled into a fighting force once landed. This all would take time, and the Army of Virginia would have to hold the fort while it was in progress.

The best defensive position for the Army of Virginia was the line of the Rappahannock River. The Rappahannock was the largest river between Washington and Richmond, and at its lower stretches was so wide that it could be defended by the navy (see map 12).

There were two Potomac River ports where McClellan's army could be landed. These

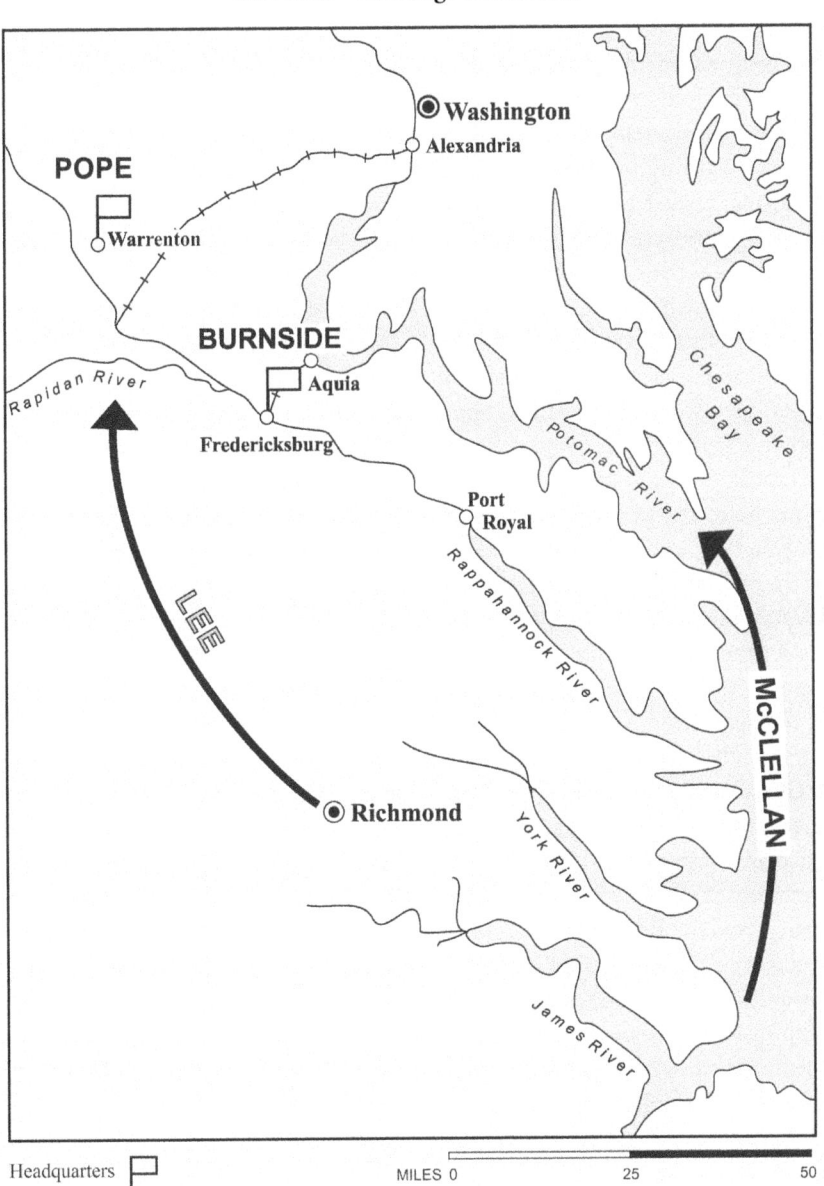

Map 12
Bull Run — Landings and Routes

were Alexandria, immediately adjacent to Washington, and Aquia, 15 miles farther down the Potomac. Both Alexandria and Aquia were rail heads on railroads that led down to the Rappahannock. The Alexandria and Orange Railroad ran from Alexandria to Rappahannock Station, and the Richmond and Potomac Railroad ran from Aquia down to Fredericksburg on the Rappahannock. Thus, if Lee attacked on the upper Rappahannock, Alexandria was the preferred landing site; and if he attacked on the central Rappahannock, Aquia was the preferred landing site.

McClellan was alerted on July 30 that his army was going to be withdrawn and was directed to immediately make arrangements for the removal of the sick and wounded. On August 3, he was formally ordered to transport his entire army to Alexandria and Aquia. As it happened, the first combat troops of the Army of the Potomac did not arrive at the Army of Virginia until August 14.

Lee was quick to recognize the vulnerability of Pope's army as the first shiploads of McClellan's army left and headed north. By the 20th, his army was lining up on the south bank of the Rappahannock in the Warrenton area opposite Pope's army on the north bank. At this point, Washington recognized that the coming showdown was going to be on the upper Rappahannock and ordered that all subsequent shiploads of McClellan's army be unloaded at Alexandria.

By August 25, the size of Pope's army was growing by leaps and bounds and the arrival of McClellan's army had changed from a trickle to a flood. Lee knew that he had to act and had to act fast. His first troops crossed the Rappahannock on August 25.

Pope's supply line was the Orange and Alexandria Railroad. It was only about 25 miles down the railroad from Alexandria to Pope's forward supply base at Manassas Junction, and only another 20 miles to the front at Rappahannock Station on the Rappahannock. Thus, it appeared that Pope's position was nearly invulnerable. At the time, trains could run at about 30 miles per hour. It was just about an hour's ride from the ships at Alexandria to his base at Manassas Junction, and less than an additional hour all the way to the front on the Rappahannock. In addition to this seemingly assured supply line, there was a telegraph line along the railroad linking Pope with both Alexandria and the War Department in Washington.

By the time the first shots were fired in the second battle of Bull Run, Pope's army had already grown to at least 63,000, with tens of thousands more bunched up at Alexandria awaiting transportation, while Lee's was static at about 54,000.[2]

However, the disparity in Pope's favor was not as great as first appears. Many of the men from McClellan's army that were rushed to Pope had only the ammunition on their persons, lacked their supply wagons and artillery, and the officers sometimes even lacked their horses.

Lee's plan of attack was daring. In fact, it went beyond daring, and many experts would call it foolhardy. The plan was to divide the army roughly in two. Jackson would take the smaller half; cross the Rappahannock; proceed north, west of the Bull Run Mountains; cross the mountains at Thoroughfare Gap; and attack and destroy Pope's forward supply base at Manassas and cut his connections with Alexandria. Jackson, instead of retreating, would then occupy a strong defensive position along an unfinished railroad cut near Manassas.

Pope could then be expected to withdraw all his troops from the Rappahannock to proceed north to attack Jackson in order to re-establish his communications with Alexandria. Once this happened, Longstreet would follow Jackson's route with the rest of the Confederate army and join Jackson.

Normally, the offensive will suffer greater casualties attacking an enemy in a prepared defensive position. By Lee's plan, which might be described as offensive-defensive although he was the attacker, the Union would be forced to attack him in a strong defensive position, and thus suffer greater casualties. Thus, Lee's plan, although seemingly risky, was really ingenious.

The Second Battle of Bull Run

We will now give an overview of the second battle of Bull Run, and then go back and see how the wording of the specific orders affected its outcome. The chronology of the battle was as follows (see map 13):

AUGUST 25

The two armies face each other on the upper Rappahannock.

AUGUST 26

Jackson takes three of Lee's eight divisions, crosses the upper Rappahannock, marches up the west side of the Bull Run Mountains, passes through Thoroughfare Gap and seizes Pope's supply base at Manassas. Pope's railroad line and communications wire to Alexandria are disrupted.

AUGUST 27–28

Pope's forces leave Rappahannock and march north to attack Jackson. Jackson does not retreat back behind the Bull Run Mountains, but takes a strong natural defense position in a railroad cut just west of Manassas. Longstreet starts north on the west side of the Bull Run Mountains.

AUGUST 29

Pope believes he still has time to destroy Jackson before Longstreet arrives, but Longstreet's troops are already pouring through Thoroughfare Gap.

AUGUST 30

Pope makes a maximum attack against Jackson, unaware that Longstreet is already on his flank. Longstreet rolls up Pope's army for a smashing Confederate victory.

The Players and the Armies

The principal players in this episode were Generals Lee, Longstreet and Jackson on the Confederate side, and Generals Pope, McDowell, and Porter on the Union side.

Map 13
Bull Run — The Second Battle of Manassas

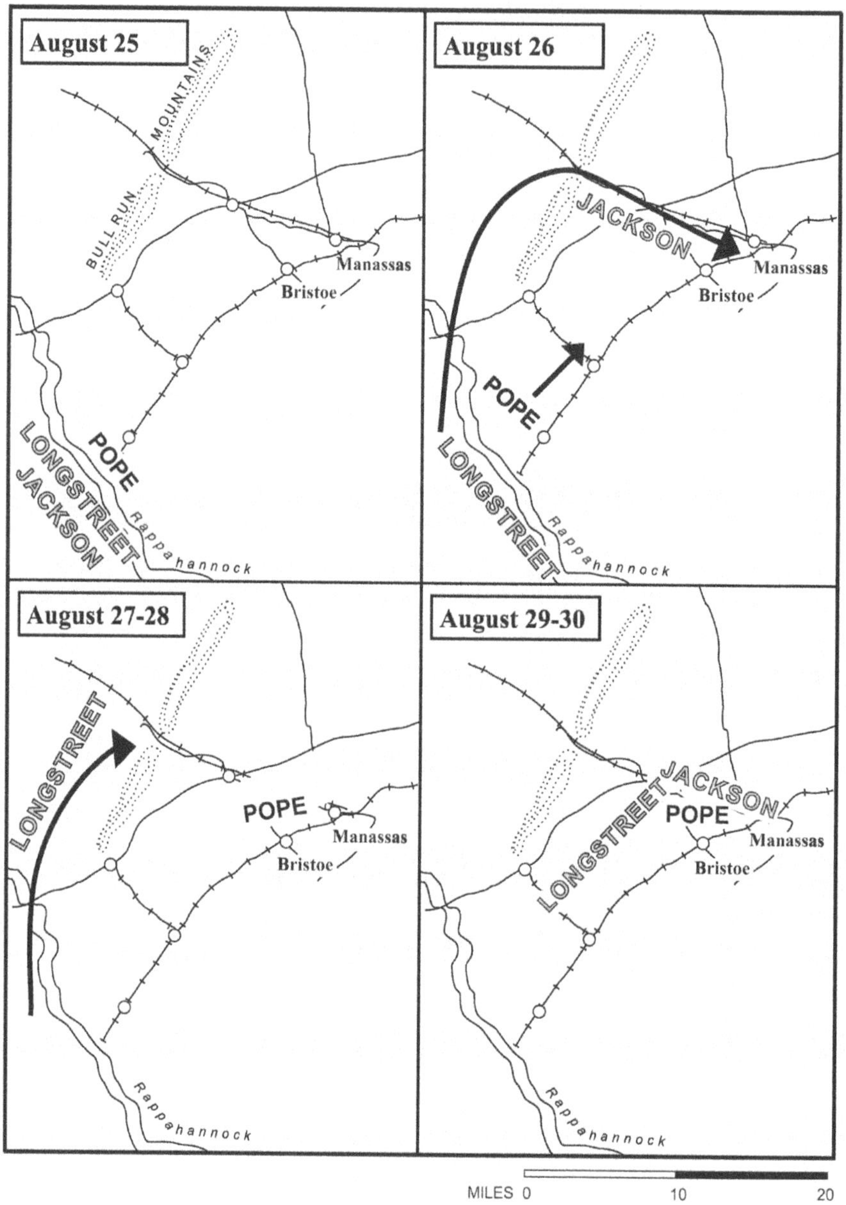

Robert E. Lee had been commanding general of the Army of Northern Virginia just 87 days when the first shots of the second battle of Bull Run were fired. During this brief period, he not only transferred the theater of operations from the gates of Richmond to the gates of Washington, but reorganized the army into a much simpler organizational structure. The army was now organized with just three commands, two corps and a cavalry command; and their three commanders reported directly to him. The corps commanders were Longstreet and Jackson, and the cavalry commander was Stuart. All three were supremely competent.

General Pope had been in command of the Army of Virginia just 60 days. However, although this was only slightly less than Lee's tenure, Pope was vastly more experienced in army command. Prior to taking command of the Army of Virginia, he had been commanding general of the Army of the Mississippi. While Lee's organizational structure had been greatly simplified, Pope's was growing more complex by the day. As we have previously noted, units from the Army of the Potomac were being rushed to him as Lee moved north. In addition, the War Department was transferring troops to him from every nearby command it could get its hands on. These various commands continued to arrive even while the battle was in progress.

By August 30, Pope had no fewer than seven corps commanders plus various lesser commanders reporting directly to him. By this time, he had been augmented with troops from the Reserve Corps in Washington, The Kanawa Division from West Virginia, the Ninth Corps from the Fredericksburg area as well as from the Army of the Potomac.

General Irvin McDowell had previously commanded the main Union army in the east and was now serving in a subordinate role as senior corps commander under Pope. McDowell had an abrasive personality and was not well liked by either his equals or his subordinates. He was, however, held in high esteem by his superiors.

Fitz-John Porter was commanding general of the Fifth Corps of the Army of the Potomac. Here he had held a privileged position. He was McClellan's good friend and closest confidant. He was also the darling of the press. He had acquitted himself well at Yorktown, Mechanicsville, Gaines Mill, and Malvern Hill, and was the nearest thing to a national hero at the time. Porter knew Pope only by reputation. He was, however, contemptuous of Pope's capabilities and resented being transferred to him. Porter reported in to Pope with his corps on August 27 after the battle had already begun. It was later contended by some that Porter gave Pope less than his best because of his attitude. Porter was subsequently charged with disobedience of Pope's orders. This case wound its way through the justice system for over 20 years before Porter was finally vindicated. However, in the eyes of some historians, he is still considered guilty.

Order Writing

Major General John Pope USV: Experienced army commander but no match for Lee.

Now let us go back to August 29, 1862. Pope had gathered his corps to the east and south of Jackson. He was determined to destroy Jackson before Longstreet could unite with him, and he believed that this could not happen before late on the

30th. In his eyes, he must destroy Jackson on the 29th or early 30th, but he was wrong. It was already too late. Longstreet's troops were already pouring through Thoroughfare Gap on the morning of the 29th.

Map 14 depicts the situation early on the 29th. General Pope's main force was concentrated to the east of Jackson along the Warrenton Pike. Porter's corps of two divisions was at Manassas Junction. Two divisions of McDowell's corps had retreated during the night. That of King was now at Manassas with Porter, and that of Ricketts was slightly below Manassas at Bristoe.

Map 14
Bull Run — Early Morning of August 29

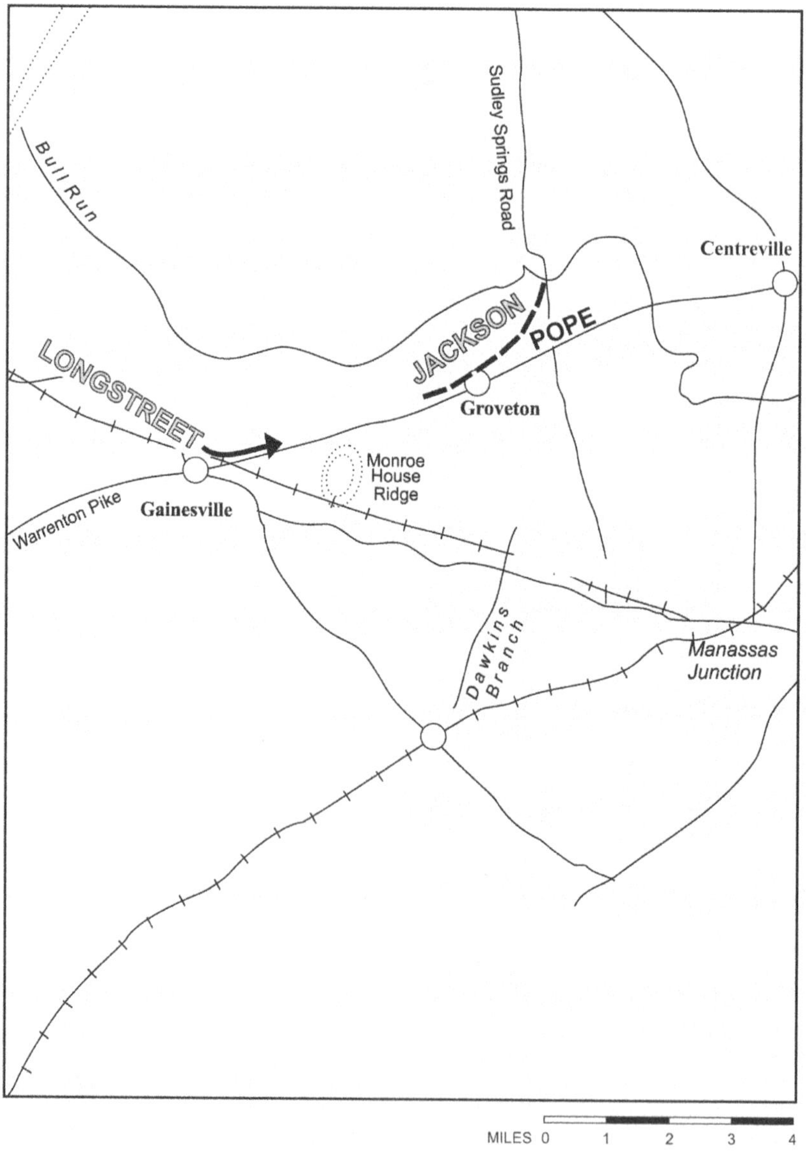

5. The Second Battle of Bull Run

A road ran from Manassas Junction to Gainesville where it intersected with the Warrenton Pike. Thus, if Pope were to continue down the Warrenton Pike, and Porter down the Gainesville-Manassas Junction Road, the two commands would ultimately meet at Gainesville. About six miles short of Gainesville, the Sudley Springs Road connected the Gainesville-Manassas Junction Road and the Warrenton Pike, and thus formed the base of an isosceles triangle with the apex at Gainesville. The distance along the Sudley Springs Road from the Gainesville-Manassas Junction Road to the Warrenton Pike was about two and a half miles (see map 14).

Corps commander McDowell had become lost during the night of August 28–29 and was not in contact with either army commander Pope or his division commanders at Manassas Junction or Bristoe. It was in this environment that Porter received the following order from Pope by courier early on the morning of the 29th:

> Headquarters Army of Virginia
> Centreville, August 29, 1862
> Push forward with your corps and King's division, which you will take with you, upon Gainesville. I am following the enemy down the Warrenton turnpike. Be expeditious or we will lose much.
> John Pope
> Major-General Commanding[3]

This order raised a number of questions as follows:

(1) Did the phrase "push forward ... upon Gainesville" mean "go to" Gainesville?
(2) What was the purpose of going down the Gainesville-Manassas Junction Road? What was Porter expected to accomplish?
(3) What did the phrase "Be expeditious or we will lose much" mean? Start soon? March fast? How far?
(4) King's division was a part of McDowell's corps. Did this order transfer King to Porter? What if McDowell showed up, who was senior to Porter?

As it happened, McDowell did show up. He met Porter in person, reported in to Pope by courier, and re-established contact with Ricketts, his division commander at Bristoe. In response to McDowell's return, Pope reissued his order to Porter. The re-issued order was greatly expanded and was addressed to both Porter and McDowell. It was as follows:

> Headquarters Army of Virginia
> Centreville, August 29, 1862
> Generals McDowell and Porter:
> You will please move forward with your joint commands toward Gainesville. I sent General Porter written orders to that effect an hour and a half ago. Heintzelman, Sigel and Reno are moving on the Warrenton turnpike, and must now be not far from Gainesville. I desire that as soon as communication is established between this force and your own, the whole command shall halt. It may be necessary to fall back behind Bull Run, at Centreville tonight. I presume it will be so on account of our supplies. I have sent no order of any description to Ricketts, and none to interfere in any way with McDowell's troops, except what I sent by his aide-de-camp last night, which were to hold his position on the Warrenton turnpike until the troops from here should fall on the enemy's flank and rear. I do not even know Ricketts' position, as I have not been able to find out where General McDowell was until a late hour this morning. General McDowell will take immediate steps to communicate with General Rick-

etts, and instruct him to join the other divisions of his corps as soon as practicable. If any considerable advantages are to be gained by departing from this order, it will not be strictly carried out. One thing must be held in view: that the troops must occupy a position from which they can reach Bull Run to-night or by morning. The indications are that the whole force of the enemy is moving in this direction at a pace that will bring them here by tomorrow night or the next day. My own headquarters will be for the present with Heintzelman's corps at this place.

John Pope
Major General Commanding[4]

This rambling order did reiterate that the column was to move toward Gainesville. However, it also introduced a number of provisions not included in the original order. These included the following: As soon as "communication is established" between the column on the Warrenton Pike and the Gainesville-Manassas Junction Road, "the whole command shall halt." The columns were now limited as to how far they could go by the provision that "the troops must occupy a position from which they can reach Bull Run by to-night or by morning." (Bull Run was behind both advancing columns.) The recipients were now given some discretion by the proviso, "If any considerable advantages are to be gained by departing from this order, it will not be strictly carried out."

The order also directed McDowell to contact his division commander, General Ricketts, and to order Ricketts's division to attach itself to the end of the column proceeding down the Gainesville-Manassas Junction Road. This McDowell did.

Finally, the order contained two items of intelligence — both wrong. It stated that the force moving down the Warrenton Pike toward Gainesville "must now be not far from Gainesville." In fact, the force was stopped by Jackson and never got farther than Groveton (see map 14). Lastly, the order stated that "the whole force of the enemy is moving in this direction at a pace that will bring them here by tomorrow night or the next day." In fact, at the very time Pope wrote his order, Longstreet's troops were flowing through Thoroughfare Gap within an hour's march from his or Porter's fronts.

This revised order contained many problems. First, it failed to say why the Porter-McDowell column was marching toward Gainesville on the Gainesville-Manassas Junction Road, or what it was supposed to accomplish. Next, the bit about not going so far that you cannot get back behind Bull Run "to-night or by morning" introduced a variability of ten miles or more. Lastly, the revised order did not specifically address the subordination of King. The original order seemingly transferred him from McDowell to Porter, but the revised order did not specifically transfer him back.

The revised order was dually addressed to McDowell and Porter. This

Major General Irvin McDowell USV: On the loser end at Bull Run 1 and Bull Run 2.

presumably indicated that Pope intended that both corps commanders operate directly under him. McDowell, however, was a man who never failed to exercise his authority. Under the so-called "rule of good discipline," under which the U.S. Armed Forces operated then and now, one must obey the last order received by a competent authority. Thus, if McDowell were to give Porter an order, even if it conflicted with the order he received from Pope, Porter would have to obey. And, as might be expected, McDowell did begin ordering Porter.

Porter dutifully started down the road to Gainesville after receiving the initial order. The order of march by division was Morell, Sykes, and King. After McDowell showed up, Ricketts joined the rear of the column. The column was now over 20,000 strong.

The head of the column passed the intersection of the Sudley Springs Road, and by about 11 A.M., reached Dawkins Branch. Dawkins Branch was a small stream that crossed the road at right angles in the middle of a cleared area. A short distance farther down the road, there was a dense wood. It was here that Porter encountered the enemy. The enemy was in the woods in unknown numbers.

Porter deployed his advance division, that of Morell, in line of battle, sent skirmishers forward, unlimbered a battery, and prepared to move forward.

At this point, McDowell rode up to the head of the column to see what the problem was. Why had the column stopped? Three separate witnesses later testified that they heard McDowell say to Porter, "Porter, you are too far out already. This is no place to fight a battle."[5] Then, after a short discussion between Porter and McDowell, McDowell rode off back down the column toward his corps. Porter then ceased his preparations for advance and took a defensive position on the near side of Dawkins Branch.

After McDowell departed, Porter was in doubt as to the subordination of General King's division in his rear. Was King still subordinate to him? Porter sent his chief-of-staff, Lt. Colonel Locke, back down the line to find out. Locke found King and McDowell in conversation at the intersection of the Sudley Springs Road. McDowell informed Locke that he was going to take King's and Ricketts's divisions and proceed up the Sudley Springs Road, leaving Porter alone on the Gainesville-Manassas Junction Road. As Locke recalled, McDowell then gave him the following verbal message for Porter:

> Give my compliments to General Porter, and say to him that I am going to the right, and will take General King with me. I think he [General Porter] better remain where he is; but if it is necessary for him to fall back, he can do so upon my left.[6]

Porter was now alone on the Gainesville-Manassas Junction Road with his two divisions, comprising about 10,000 men. His forward division, that of Morell, was in a defensive position at right angles to the road on the near side of Dawkins Branch. His other division, that of Sykes, was still in column behind with stacked arms, lounging alongside the road. Directly ahead of Morell, in the woods, was the enemy. According to reports of Porter's skirmishers, and by virtue of clouds of dust he could see with his own eyes that extended all the way from the woods back to Thoroughfare Gap, the enemy force was very large and rapidly growing. The enemy was Longstreet.

Porter had not been able to confer with Pope since early on the preceding day. The only instructions he had or was to receive until near dark on the 29th are the ones we have seen. These were:

From Pope: (1) You will push forward with your command toward Gainesville; (2) As soon as communications are established between the two columns (Porter's and Pope's) both will stop; (3) You will not go so far that you cannot get back to Bull Run tonight or tomorrow.

From McDowell: (1) Porter, you are too far out already. This is no place for a battle; (2) He (Porter) better remain where he is, but if he must fall back, he should do so on my (McDowell's) left.

As of the time Locke returned to Porter with McDowell's message, communications had not been established between the two columns and both were stopped; Pope's by Jackson at Groveton, and Porter's by the force in the woods beyond Dawkins Branch.

Porter could not see Pope's column because of the dense woods between them, and he did not know what the distance was between the two stopped columns. Actually, it was a little over two miles. Porter could, however, hear the sounds of battle wafting through the woods as Pope fought Jackson. And so, as Pope fought Jackson, Porter's corps, within sound of the battle, remained idle.

As the day progressed, Pope, over on the Warrenton Pike, was putting ever more pressure on Jackson. By midafternoon, things seemed to be going Pope's way, and Pope kept expecting Porter to momentarily appear on Jackson's flank or rear and administer the coup de grace; but Porter did not appear. Finally, in exasperation, at 4:30 P.M. Pope wrote out a peremptory order to Porter for Porter to attack Jackson immediately. Pope gave the order to a courier for delivery and then waited — and waited, and waited. Porter's headquarters was just about three miles from that of Pope, so the order should have been delivered within a half an hour. The battlefield order was as follows:

Headquarters in the Field
August 29, 4:30 P.M.
Major-General Porter:
   Your line of march brings you in on the enemy's right flank. I desire you to push forward into action at once on the enemy's flank, and if possible, on his rear, keeping your right in communication with General Reynolds. The enemy is massed in the woods in front of us, but can be shelled out as soon as you engage their flank. Keep heavy reserves and use your batteries, keeping well closed to your right all the time. In case you are obliged to fall back, do so to your right and rear, so as to keep you in close communication with the right wing.
   John Pope
   Commanding General[7]

Unknown to General Pope, the courier got lost and spent over three hours looking for Porter. When Porter finally received the order, it was too dark to implement.

Darkness came and Jackson survived. Pope was furious. He felt sure that Porter had disobeyed his order. He now drafted the following order and handed it to a courier:

Headquarters Army of Virginia
In the field, near Bull Run
August 29, 1862, 8:50 P.M.
Maj. Gen. F. J. Porter:
   General: Immediately upon receipt of this order, the precise hour of receiving which you will acknowledge, you will march your command to the field of battle of to-day and report to me in person for orders. You are to understand that you are expected to comply strictly with this order, and to be present on the field within three hours after its reception, or after daybreak to-morrow morning.
   John Pope
   Major-General Commanding[8]

Porter did not receive the order until after midnight, but reacted to it immediately. At the time he received the order he was with Sykes's division, which was still strung out along the Gainesville-Manassas Junction Road behind Morell, just as it was at 11 A.M. that morning. Porter immediately orally ordered Sykes to take his division and proceed up the Sudley Springs Road toward the Warrenton Pike, where the battle had been raging that day until dark. Two of Porter's units were beyond verbal command. Morell's division was still deployed ahead, behind Dawkins Branch, and as Porter started down the road toward Gainesville that morning, he left the independent brigade of Sturgis behind in Manassas Junction as a reserve and to serve as a rallying point in case he was compelled to retire. Porter thus prepared written orders for Morell and Sturgis and handed them to couriers for delivery. The orders were as follows:

> General Morell:
> Lose not a moment in withdrawing and coming down the road to me. The wagons which went up send down at once and have the road cleared, and send me word when you have all in motion. Your command must follow Sykes.
> F. J. Porter
> Commanding[9]

> General Sturgis:
> Please put your command in motion to follow Sykes as soon as he starts. If you know of any other troops who are to join me I wish you to send notice to them to follow you.
> F. J. Porter
> Major-General[10]

Note that neither the order to Morell nor that to Sturgis told them where to go. Both were merely told to follow Sykes. At the time the orders were disseminated, Sykes was on the Gainesville-Manassas Junction Road slightly to the west of the intersection with the Sudley Springs Road, while Sturgis was on the road almost two miles to the east of the intersection. Inasmuch as Sykes turned up the Sudley Springs Road, he never passed Sturgis, and Sturgis waited in vain to follow him (see map 15).

When Morell received his order, he dutifully passed on the order to his three brigade commanders and two of the three correctly put their commands on the road to follow Sykes. However, Morell's skirmishers, who were deployed beyond Dawkins Branch in the woods, belonged to the brigade of Griffin, and Griffin could not take to the road until they returned. By the time they returned and Griffin was ready to take to the road, the Sykes column was out of sight.

Griffin marched down the road and continued right past the Sudley Springs intersection, having no reason to turn off. Griffin ultimately passed Sturgis's troops, who were still waiting in Manassas Junction for Sykes to pass. Sturgis, seeing a column of Porter's troops march past, fell in behind. The two commands marched all the way to Centreville, away from the battle, before learning of their error (see map 15).

The brigade of Sturgis did ultimately reach the battlefield in the late afternoon to participate in the final phase of the fighting, but that of Griffin never reached the battlefield. Sturgis's brigade was unusually large, numbering about 3,500 men. Griffin's contained about 1,500. In addition, each had one battery of artillery. In consequence, Pope was deprived of 5,000 men and two batteries of artillery for most of the day's fighting, and 1,500 men plus one battery of artillery for all of the day's fighting.

Map 15
Bull Run — Griffin's March of August 30

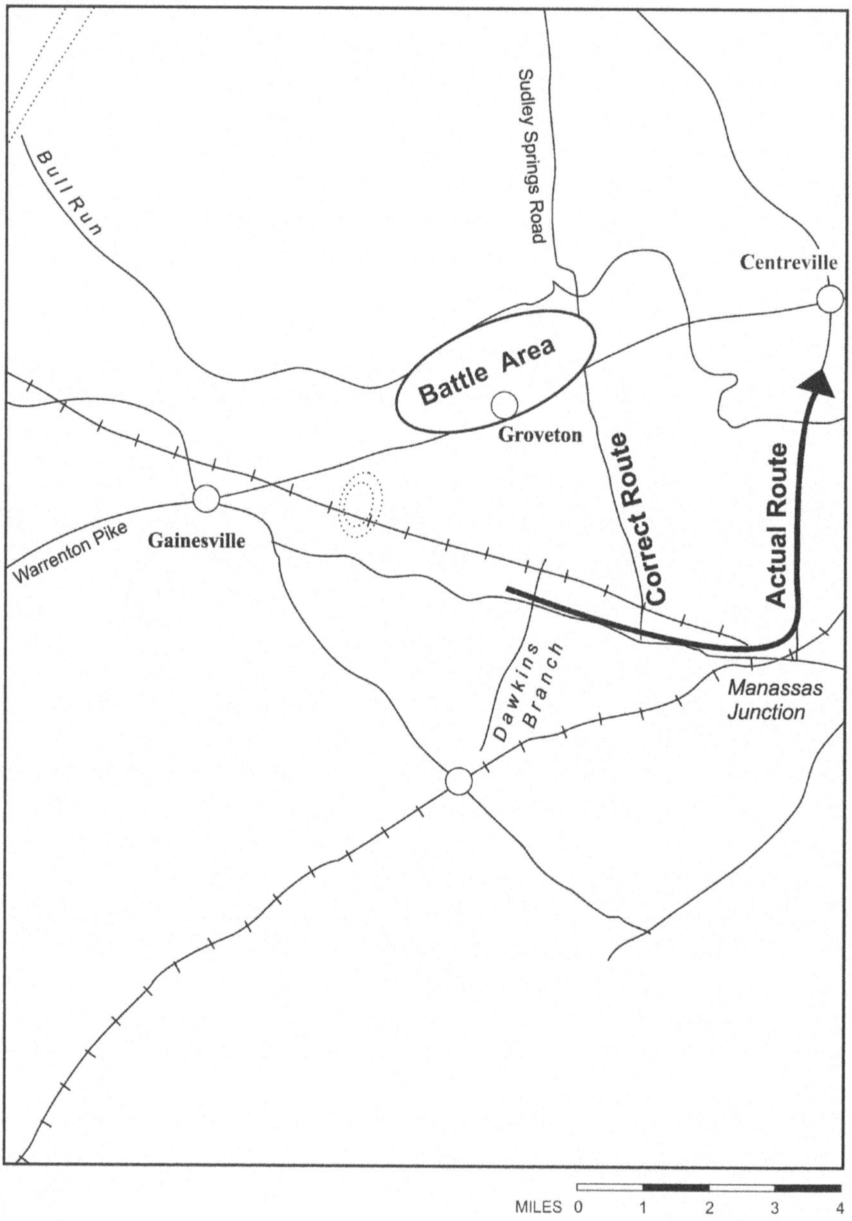

Now let us sum up and see if we can evaluate the effect that bad order writing had on the outcome of the battle. Pope's bad order writing was at least partially responsible for Porter's 10,000 sitting idle within the sound of the guns on August 29. Until Pope's order of 4:30 P.M., which Porter did not receive until too late to implement, he never told Porter what he wanted Porter to accomplish. However, as it turned out, Porter's idleness was not all bad. His presence at Dawkins Branch tied down the large force of Longstreet's troops facing him that might have otherwise been thrown into Jackson's fight with Pope. Furthermore, had not Porter been on the Gainesville-Manassas Junction Road at Dawkins Branch, Longstreet could have done to Pope what Pope had hoped to do to Jackson — that is, hit him in the flank and rear. Thus, we may conclude that Pope's bad order writing did not necessarily contribute to his defeat.

We cannot say the same thing about Porter's bad order writing. As a result of his bad order writing, Pope definitely lost the full services of Griffin and almost the full services of Sturgis on the crucial 30th. If we ask the question, did this have an effect on the outcome of the battle?, the answer must be "yes." If we ask the further question, was the effect enough to change the outcome of the battle?, we must answer that one will never know.

Let us close by quoting Pope's answer to the question as to the effect on the outcome:

> A very great effect. I do not know the strength of General Griffin's brigade; but a brigade of four regiments and a battery of artillery, as I understand it. That was utterly withdrawn from the field; took no part in the action. General Piatt's command [Sturgis] got up very late; too late to do anything, except, indeed to contribute to enable us to maintain our ground until darkness closed the fight. The presence of the other brigade would undoubtedly have been of immense benefit.[11]

6

The Maryland Campaign of September 1862

Let us start by giving an overview of what transpired during the Maryland Campaign of September 3 to 18 of 1862. We will then go back and examine how battlefield orders impacted on the campaign.

On August 30, the Confederate army of Robert E. Lee inflicted a smashing defeat on the Union army of John Pope, in the second battle of Bull Run. It appeared that even Washington was lost. In this crisis, President Lincoln put General George B. McClellan in charge of all the troops in the Washington area. McClellan quickly created order out of confusion and, in a matter of days, this rejuvenated army was ready to take to the field again, more formidable than ever. For the rest of the campaign, it was the Confederate army of Lee versus the Union army of McClellan.

Lee, after his victory, decided to invade the North and carry the conflict into Union territory. On September 3, Lee's army crossed the Potomac at Leesburg and headed north. However, as Lee's army advanced and McClellan's reorganized, the Confederate army began to melt away through straggling as the Union army gained strength. This growing disparity was not, however, evident to McClellan, who continued to believe that the Confederate army was more numerous than his. The actual strength of Lee's army at this time was about 40,000, while McClellan's was over 80,000. Lee's objective was Harrisburg, Pennsylvania. McClellan thought it was Washington, Baltimore, or perhaps Philadelphia.

During the first days of the invasion, Lee's army advanced northward to Frederick, Maryland, while McClellan's army slowly moved out of Washington toward Frederick to maintain a blocking position between the Confederate army and Washington and Baltimore. During this phase, there were few casualties and only clashes of small detachments of cavalry that were operating between the armies. Up to now, mountains played no part in the movements of the armies. This was all to change on September 9–10. Frederick, Maryland, lay at the foothills of the Appalachian Mountains. One could not go beyond Frederick without hitting the mountains.

The Appalachian Mountains are not what one often pictures mountains to be. They are not conical rises. Rather, they are continuous, long, high ridges generally running from north to south. The ridges extend as far north as Harrisburg, Pennsylvania, and as far south as central Virginia. These ridges are only infrequently broken by passes or river valleys, and it is these one must use to cross the ridges. Throughout most of the Appalachian range, there are at least two parallel ridges with a fertile valley in between. The valley north of the Potomac is called the Cumberland Valley, and south of the Potomac it is called the Shenandoah Valley.

6. The Maryland Campaign of September 1862

Upon reaching Frederick on September 9, Lee decided to move his further advance and trailing supply line over the ridge into the valley. He thus felt his supply line would be more secure as there were relatively few passes into the valley.

McClellan had been moving slowly and cautiously westward since the invasion started, at a rate of about six miles a day. By September 10, he stood before the ridge blocking his further progress. In order to proceed further, he had to force the two passes before him. These were Turner's and Fox's Gaps; Turner's was directly before him on the National Road, and Fox's one mile to the south.

To proceed beyond this point was fraught with danger. In his eyes, a numerically superior enemy held the high ground. He would need more information before proceeding.

As Lee's army prepared to cross over the mountain and proceed up the valley, they encountered a new problem. There were two Union garrisons in their rear, blocking their supply route into Virginia. These were the garrisons at Harpers Ferry and Martinsburg. Both were on the south side of the Potomac; Harpers Ferry on the east side of the valley and Martinsburg on the west side of the valley. The distance between the two is about 15 miles. The larger garrison consisting of about 12,000 was at Harpers Ferry and was commanded by Colonel Miles. The smaller garrison of about 2,000 at Martinsburg was commanded by Brigadier General White. The reason we have a colonel commanding the large garrison and a general the small is because the army high command had greater confidence in the colonel. He was a professional officer with over 30 years experience, wherein White was a recent political appointee.

Major General George B. McClellan USV: Classmate of Jackson, a cautious man.

On September 9, just before moving the Confederate army into the valley, Lee and Stonewall Jackson came up with a daring plan to remove the Union garrisons. The plan called not only for their removal, but the capture of the lot. Some have called the plan ingenious, many others foolhardy. Lee and Jackson would never have chanced it had they both not known McClellan personally and well. Jackson had been a four-year classmate of McClellan at West Point, class of 1846. Lee had served jointly with McClellan on General Scott's staff during the Mexican War. Both believed they could hoodwink him because of his excessive caution.

The plan was "Special Order 191." It was to be executed beginning September 10. Jackson would take three of Lee's nine divisions, cross the Potomac, make a wide swing to the west, and drive the Martinsburg garrison into Harpers Ferry. General McLaws would take two divisions and approach Harpers Ferry from the north. He would seize the vital Maryland Heights that looked down on Harpers Ferry from the north side of the Potomac. General Walker would cross the Potomac with one division and box in Harpers Ferry from the east. The three groups planned on meeting and assuming their positions surrounding Harpers Ferry on Friday, September 12. Jackson, who was senior of the three generals, and who was a former commanding officer of Harpers Ferry and thus knew the terrain intimately, would then take charge and complete the capture of the surrounded garrison.

In the meantime, the army would leave its wagon supply train and artillery reserve parked at Boonsboro. Two divisions of Longstreet would remain at Boonsboro, and the last of the nine divisions, that of D. H. Hill, would guard the passes in the mountains separating them from McClellan's army (see map 16).

The plan visualized the quick capture of Harpers Ferry and the beginning of the reuniting of the army at Boonsboro by September 13. However, as of the 13th, the plan was way behind schedule. Jackson only arrived at the outskirts of Harpers Ferry at noon on the 13th, and as of late in the day was still trying to coordinate operations with the other two forces. The battle of Harpers Ferry had not even begun.

The plan then took another turn for the worse. The two divisions of Longstreet had to be dispatched 20 miles to the west to Hagerstown to break up a gathering of militia. Now General Hill's division stood alone in the passes between McClellan's vast army and the vulnerable wagon train and artillery park at Boonsboro. If McClellan made a quick and resolute attack, there was absolutely no doubt he could push Hill aside and seize the wagons and artillery park. If McClellan made a slower and more methodical attack, perhaps there would be time for Longstreet to return in time to help Hill. Perhaps they could then hold McClellan off long enough for the wagons and artillery to escape, and then withdraw to the nearest place where they could unite with the forces besieging Harpers Ferry. That place was Sharpsburg (Antietam Creek). Sharpsburg was slightly north of the Potomac. Behind it was a ford in the river where the Confederate besieging units south of the Potomac could cross and reunite.

Colonel Dixon Miles USA: Commanding Officer at Harpers Ferry who unknowingly held the keys to victory.

There was little doubt that Longstreet's two divisions and Hill's division could retire to Sharpsburg after defending the passes, but what about the six divisions besieging Harpers Ferry? Here we had a variable. It depended on how long the Union garrison held out. Whether or not it held out, Jackson, whose force was on the Sharpsburg side of the siege, could always retire to the reunite point at Sharpsburg. Not so for Walker, who was to the east of Harpers Ferry. If the garrison surrendered, Walker could simply march through Harpers Ferry and follow Jackson to Sharpsburg. If the garrison held out, however, Walker was isolated and could not reach Sharpsburg in time for the pending battle.

The case of the two divisions of McLaws, which were on the north side of the Potomac

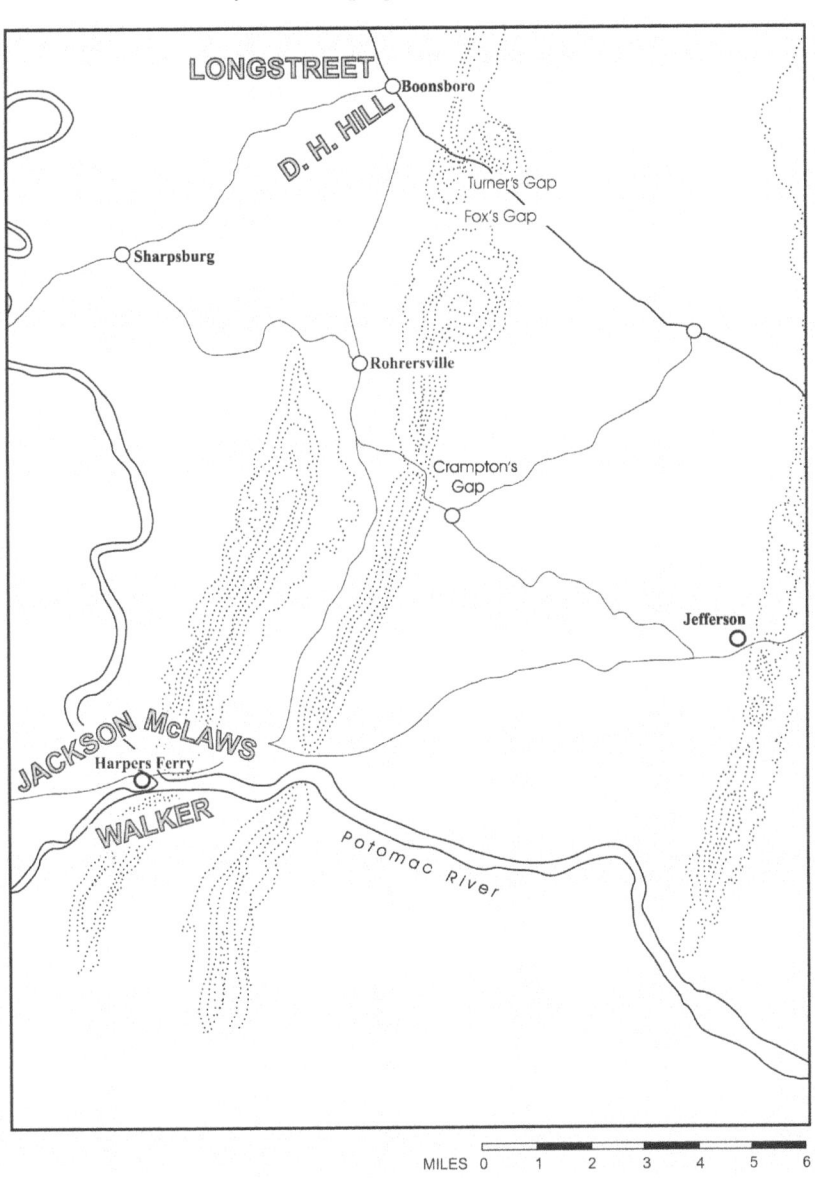

Map 16
Maryland Campaign — Special Order 191

opposite Harpers Ferry, was more problematic. If Harpers Ferry surrendered, he could simply cross the bridge into Harpers Ferry and follow Jackson and Walker to Sharpsburg. If the garrison did not surrender, however, he had only one possible route to Sharpsburg. He had to march up the Pleasant Valley road to a place call Rohrersville and then turn to his left to take the road that lead to Sharpsburg (see map 17). However, Rohrersville was just two miles below the passes where McClellan would break through. When and if he did break through, he could be expected to quickly close off the road from Rohrersville to Sharpsburg. McLaws then could not join Lee at Sharpsburg for the pending showdown battle.

We can see that Harpers Ferry was the key to the whole situation (see map 17). If McClellan attacked and Harpers Ferry held out, the full Confederate army could not quickly reunite at Sharpsburg in time for the showdown battle. McClellan would destroy the Confederate army in detail. It all depended on the Union garrison at Harpers Ferry.

As of September 13, the Union commanders at Harpers Ferry were in near despair. They had been cut off from all communications with the outside world since the 11th. First, they were aware of vast Confederate forces to their east, then to their north. Now they were completely surrounded by what they considered to be overwhelming Confederate forces. They had no knowledge of any relief effort. They had no knowledge of the location of any Union force. They had known nothing but Union defeat since the second battle of Bull Run. The Confederates looked invincible and morale at Harpers Ferry was low.

It was in this environment that a fantastic, almost unbelievable thing happened. On the morning of Saturday, September 13, a Union corporal picked up three cigars that were wrapped in a paper. The paper was a copy of the Confederate secret battle plan, Special Order 191. By early afternoon, a copy was before General McClellan. Upon reading the plan, McClellan stated, "Here is a piece of paper with which if I cannot whip Bobby Lee, I will be willing to go home."[1] He now saw the dispersal of the Confederate army before him in all detail. He now had the unprecedented opportunity to destroy it in detail. As McClellan read the plan, his situation was as follows (see map 17): Four of his six corps were bunched up near Frederick at the base of South Mountain, a fifth was en route to join him. Directly in front of McClellan was Turner's Gap and one mile south of Turner's Gap was Fox's Gap. Just on the other side of the mountain was Boonsboro, with the Confederate wagon train and artillery reserve.

Ten miles to the south of McClellan at Jefferson was McClellan's other corps, headed by General Franklin. Franklin had two divi-

Major General Lafayette McLaws CSA: Besieger and besieged.

Map 17
Maryland Campaign — McClellan's Situation on September 13

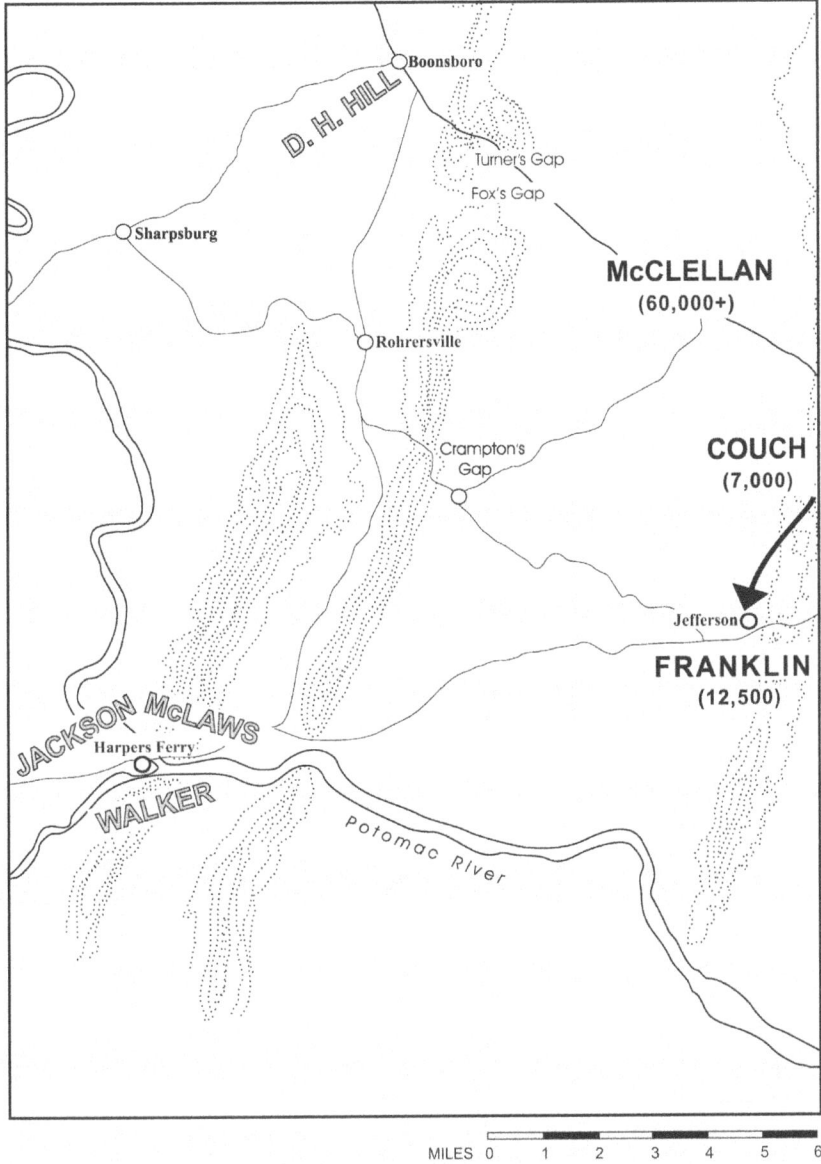

sions with him totaling 12,500 men. In addition, he had a third division of 7,000 men coming up eight hours behind. Franklin thus headed a potent force of almost 20,000 men. This force was nearest to Harpers Ferry. If Harpers Ferry was to be relieved, this force would have to do it.

According to the paper before McClellan, only two of the six divisions of Confederate troops besieging Harpers Ferry were north of the Potomac. These were the two divisions of McLaws. These two, if attacked, could not be quickly reinforced by the others as the Union garrison controlled the only bridge. McClellan did not know the strengths of

McLaws's two divisions, but it was actually about 8,000. Thus, if so tasked, Franklin with his 20,000 should have been easily able to beat his way into Harpers Ferry.

There were two routes by which Franklin could reach the garrison at Harpers Ferry (see map 17). The more direct route was to continue to follow the main road from Jefferson southwest to the Potomac. Here, the road curved to the west, passed between the base of South Mountain and the river, and continued on to the bridge at Harpers Ferry. The distance from Jefferson to the Potomac was about nine miles. The distance from the pass at the base of South Mountain to the bridge at Harpers Ferry was another two and a half miles. It could be expected that the Confederates would defend the pass between the mountain and the river.

The other route from Jefferson to Harpers Ferry was longer and more circuitous. In this route, one had to proceed northwest from Jefferson to Crampton's Gap, pass through the gap, and proceed eight miles down Pleasant Valley to the bridge. In this instance, it could be anticipated that the Confederates would defend Crampton's Gap.

It would appear that the two approaches might be likened to "six of one and half dozen of the other." In either event, the Union force would have to pass through a defensible stricture point to reach its objective: in the case of the river approach, the gap between the base of South Mountain and the river; in the case of the other approach, Crampton's Gap.

The river approach, however, had a huge advantage. Here, the stricture point, the point where conflict would begin, was just two and half miles from the garrison. The beleaguered garrison would be subject to the sights and sound of battle. They would know that relief was at hand. In the case of the Crampton's Gap approach, the stricture point, and hence location of battle, was eight miles plus from the garrison. As we subsequently learned, they never heard nor saw signs of relief. Sound propagation in mountains is capricious. Along the river, it is unimpeded.

The preferential approach was the river approach. This was recognized by the Confederates. During the morning of the 14th, Confederate cavalry commander, General Stuart, visited Crampton's Gap and found that two of his brigades were camped there. He immediately directed General Wade Hampton to move his brigade to cover the river gap.[2] During the day of the 13th, General Lee sent a dispatch to General McLaws warning him to watch for the approach of Federals via the river approach.[3]

Now let us return to General McClellan, who had just read and digested Special Order 191 and was ready to issue his orders. He ordered the four corps that were with him to deploy for and attack Fox's and Turner's Gaps at daybreak on the 14th. He then prepared his order for General Franklin. The order bore a time of 6:20 P.M. and was turned over to a courier for delivery. We don't know the time of delivery, but it was probably late at night as the 13th turned over into the 14th. The order was as follows:

HEADQUARTERS ARMY OF THE POTOMAC
Camp near Frederick, September 13, 1862 — 6:30 P.M.
Maj. Gen. W. B. FRANKLIN,
   Commanding Sixth Corps:
   GENERAL: I have now full information as to movements and intentions of the enemy. Jackson has crossed the Upper Potomac to capture the garrison at Martinsburg and cut off Miles' retreat toward the west. A division on the south side of the Potomac was to carry Loudon Heights and cut off his retreat in that direction. McLaws, with his own command and the division of R. H. Anderson, was to move by Boonsborough and Rohrersville to carry

the Maryland Heights. The signal officers inform me that he is now in Pleasant Valley. The firing shows that Miles still holds out. Longstreet was to move to Boonsborough and there halt with the reserve corps, D. H. Hill to form the rear guard, Stuart's cavalry to bring up the stragglers, &c. We have cleared out all the cavalry this side of the mountains and north of us.

The last I heard from Pleasonton he occupied Middletown, after several sharp skirmishes. A division of Burnside's command started several hours ago to support him. The whole of Burnside's command, including Hooker's corps march this evening and early tomorrow morning followed by the corps of Sumner and Banks and Sykes' division, upon Boonsborough, to carry that position. Couch has been ordered to concentrate his division and join you as rapidly as possible. Without waiting for the whole of that division to join, you will move at daybreak in the morning, by Jefferson and Burkittsville, upon the road to Rohrersville. I have reliable information that the mountain pass by this road is practicable for artillery and wagons. If this pass is not occupied by the enemy in force, seize it as soon as practicable, and debouch upon Rohrersville, in order to cut off the retreat of or destroy McLaws' command. If you find this pass held by the enemy in large force, make all your dispositions for the attack, and commence it about half an hour after you hear severe firing at the pass on the Hagerstown pike, where the main body will attack. Having gained the pass, your duty will be first to cut off, destroy, or capture McLaws' command and relieve Colonel Miles. If you effect this, you will order him to join you at once with all his disposable troops, first destroying the bridges over the Potomac, if not already done, and, leaving sufficient garrison to prevent the enemy from passing the ford, you will then return by Rohrersville on the direct road to Boonsborough if the main column has not succeeded in its attack. If it has succeeded, take the road by Rohrersville to Sharpsburg and Williamsport, in order either to cut off the retreat of Hill and Longstreet, toward the Potomac, or prevent the repassage of Jackson. My general idea is to cut the enemy in two and beat him in detail. I believe I have sufficiently explained my intentions. I ask of you, at this important moment, all your intellect and the utmost activity that a general can exercise.

GEO. B. McCLELLAN
Major-General, Commanding.[4]

The wording of this order probably determined the future of the war. Let us see how it all actually played out and then we will go back and see how it might have played out had the wording of the order been different.

There were actually two separate and distinct battles on September 14, even though they were only five miles apart. These were McClellan's battle of Turner's and Fox's Gaps, and Franklin's battle at Crampton's Gap. Often, in historical accounts, the two battles are lumped together under the single title, "The Battle of South Mountain." Let us look at McClellan's battle first.

Instead of making a rushing attack early in the day, which undoubtedly would have carried the gaps, McClellan made a slow, deliberate, carefully organized attack that did not pick up steam until the afternoon. This gave Longstreet's two divisions at Hagerstown time to cover the 20 miles back to the gaps and assist General Hill. McClellan's forces did not penetrate the gaps until after dark. By this time, the wagon park and artillery reserve at Boonsboro had been safely withdrawn, and the defending divisions of Longstreet and that of Hill were safely on the road to Sharpsburg for their subsequent reuniting with the Harpers Ferry siege forces.

Now to Franklin's battle at Crampton's Gap. In accordance with his instructions, Franklin marched his corps to Burkittsville and then to the base of Crampton's Gap, arriving at about noon. He then spent three hours reconnoitering the tiny Confederate defense force, which he outnumbered at least ten to one. He finally attacked and the gap was his by dark.

However, he still had eight miles to go down Pleasant Valley to reach the Harpers Ferry bridge. He waited until the morning of the 15th to resume his attack. By this time, it was too late. The Harpers Ferry garrison, knowing nothing of his relief attempt, surrendered sometime after 8 A.M. Worse yet, McLaws's two Confederate divisions, which were supposed to be trapped between Franklin and the Harpers Ferry garrison, were now free to retreat across the bridge and through Harpers Ferry.

On the 15th, the day following the battles of the gaps, McClellan declared it a huge Union victory. He had "shockingly whipped Bobby Lee."[5] But had he? In his northern battle, he really had two objectives: first, to destroy or capture the wagon park and artillery reserve; and second, to destroy the defending Confederate forces. He did neither.

Now to Franklin's battle. According to his orders, he too had two objectives, and we will quote the order, "Your duty will be first to cut off, destroy, or capture McLaws command and relieve Colonel Miles." Franklin accomplished neither.

Now let us look at the comparative casualties for the 48 hours following the time McClellan sat down to write his battle orders on the 13th.

	Union Losses	*Confederate Losses*
Combined battles at South Mountain	2,325	3,000 (approx)
Harpers Ferry	12,737	250 (approx)
Total	15,062	3,250 (approx)

Thus, the Federals suffered over four and a half times the number of casualties as the Confederates and achieved none of their objectives, while the Confederates achieved all of theirs. Namely, they held Fox's and Turner's Gaps long enough for the evacuation of the wagon train and artillery reserve and for the defending forces to successfully withdraw to Sharpsburg and they held Crampton's Gap long enough for the completion of the capture of Harpers Ferry and the concomitant escape of McLaws. In addition, the Confederates captured enough cannon, rifles, and equipment at Harpers Ferry to outfit an entire corps. Some Union victory.

We can see from the above that Franklin's potent force of almost 20,000 men was entirely wasted on that crucial day of September 14, 1862. The operations on this date determined whether or not the Confederates could unite for the Battle of Antietam, and how quickly.

From the morning of the 15th onward, there was nothing but time and distance to prevent the uniting, and unite they did. Jackson's troops arrived at Sharpsburg on the morning of the 16th, Walker's on the afternoon of the 16th, McLaws's on the morning of the 17th when the battle was already in progress, and A. P. Hill's division, in most dramatic fashion, on the afternoon of the 17th. On the afternoon of the 17th, it appeared that Confederate defeat was certain and the destruction and loss of the Confederate army at hand. Burnside's corps was advancing towards Lee's rear, and Lee had no more reserves to stop him. At this instant, a column of troops appeared in the distance advancing from the direction of Harpers Ferry. Lee nervously looked at them through his binoculars. If Union, all was lost. But they were not Union; they were A. P. Hill's 4,000-man Confederate division, the last piece of Lee's army to unite. Hill turned the tables on the advancing Union troops and the battle ended in a draw. It was that close.

Now let us go back and look at McClellan's order of the 13th to Franklin and see if it all might have turned out differently had the order been otherwise. The order was long, complex, called for multiple objectives that were buried in the text, and was replete with detailed instructions as to how the objectives were to be accomplished.

Some objectives were such that McClellan himself, with uncommitted troops available to him, could accomplish as easily or more easily than Franklin. For example, "Debouch upon Rohrersville to cut off the retreat of or destroy McLaws' command." Rohrersville was about two miles south of Fox's Gap, which McClellan was to seize, and about two miles north of Crampton's Gap, which Franklin was to seize, and on the same valley road. Thus, McClellan should have been able to reach Rohrersville just about as quickly as Franklin.

There was one thing that only Franklin could do and not McClellan. That was the expeditious relief of the garrison at Harpers Ferry. Franklin was ten miles closer than McClellan. Despite this, the "expeditious relief" of Harpers Ferry was not highlighted in the order to Franklin.

Let us know recast McClellan's order of the 13th to Franklin and hypothesize if it would have made a difference:

HEADQUARTERS ARMY OF THE POTOMAC
Camp near Frederick, September 13, 1862 — 6:30 P.M.
Maj. Gen. W. B. FRANKLIN,
  Commanding Sixth Corps:
GENERAL: I have firm information on the disposition of Confederate forces. Six divisions are besieging Harpers Ferry and three are before me in the Boonsboro area.

Of the besieging forces, Jackson with three division is south of the Potomac and west of Harpers Ferry. Walker with one division is south of the Potomac and east of the city. McLaws with two divisions is north of Harpers Ferry and on this side of the Potomac. Thus, in any attempt to reach the bridge at Harpers Ferry and relieve the garrison, you will only have to deal with McLaws's two divisions and with whatever cavalry is in the area.

You are hereby tasked with the expeditious relief of Harpers Ferry. Time is of the essence. This is already one day later than the Confederates anticipated the surrender of the garrison. You must not waste one hour.

My intention is to pass through Turner's and Fox's Gaps and destroy the Confederates in the Boonsboro area. I will attack at dawn tomorrow. In addition to the above, once through the gaps, I will occupy Rohrersville, thus cutting McLaws off from retreating to Sharpsburg.

Keep me informed of your actions and intentions by hourly reports.
George B. McClellan
Major General Commanding

Under the above, Franklin would have a single, clearly defined task, and he could choose the means and routes to accomplish it. We will, of course, never know how it would have come out. Could Franklin, with his 19,000 plus men have prevailed over McLaws's 8,000 in time to save the garrison? Actually, Franklin need not have had to relieve the garrison to give McClellan a winning hand at Antietam three days hence. He need only have succeeded in stiffening the will of the garrison to hold out a little longer than it did.

General-in-chief of all the Union armies, Henry Halleck, expressed an opinion on these events during testimony he gave before the Harpers Ferry Military Commission on October 29, 1862. Halleck said, "I am of the opinion that it was possible for General McClellan to have relieved and protected Harpers Ferry and that he should have done so."[6]

Let us here hypothesize a different ending for the battle of Antietam. It is the afternoon of September 17, and everything is as it actually was. Burnside is advancing toward the

Confederate rear. Lee has used his last reserves. Certain defeat and destruction seem inevitable. Then, Lee sees in the distance a column advancing from the direction of Harpers Ferry. Is it Union or Confederate? It is the 12,700-man Union garrison from Harpers Ferry that Franklin relieved on the 14th.

Before we leave the Maryland campaign, let us take a brief look at the battlefield order writing capabilities of the Confederates.

On September 9, 1862, Lee and Jackson agreed upon a plan for capturing the Union garrisons at Martinsburg and Harpers Ferry in their rear. Walker, with one division, was to take a blocking position to the east of Harpers Ferry. McLaws, with two divisions, was to capture the vital Maryland Heights that dominated Harpers Ferry from the north, and then take a blocking position to the north. Jackson, with three divisions, was to make a wide sweep to the west, drive the Martinsburg garrison into Harpers Ferry, and then take a blocking position to the west of Harpers Ferry. The encirclement was to be completed on September 12. Jackson was then to take command and complete the capture of the beleaguered garrison.

We know this was the intent from a variety of sources. These include the actions of the participants, the dispatches exchanged during the operation, and the post operation reports.

After Lee and Jackson separated, the plan was reduced to writing and then delivered by courier to each of the players. They had no further opportunity for discussion. The written confirmation, designated "Special Order 191," was signed by Lee's chief-of-staff, Colonel Chilton.

Did Special Order 191 really validate what Lee and Jackson had agreed upon? It did not! Following is Special Order 191 verbatim:

Special Orders, No. 191
Headquarters Army of Northern Virginia
September 9, 1862

1. The citizens of Fredericktown being unwilling while overrun by members of this army, to open their stores, in order to give them confidence, and to secure to officers and purchasing supplies for benefit of this command, all officers and men of this army are strictly prohibited from visiting Fredericktown except on business, in which cases they will bear evidence of this in writing from division commanders. The provost marshal in Fredericktown will see that this guard rigidly enforces this order.

2. Major Taylor will proceed to Leesburg, Va., and arrange for transportation of the sick and those unable to walk to Winchester, securing the transportation of the country for this purpose. The route between this and Culpepper Court-House east of the mountains being unsafe, will no longer be traveled. Those on the way to this army already across the river will move up promptly; all others will proceed to Winchester collectively and under the command of officers, at which point, being the general depot of this army, its movements will be known and instructions given by commanding officer regulating further movements.

3. The army will resume its march tomorrow, taking the Hagerstown road. General Jackson's command will form the advance, and, after passing Middletown, with such portion as he may select, take the route toward Sharpsburg, cross the Potomac at the most convenient point, and by Friday morning take possession of the Baltimore and Ohio Railroad, capture such of them as may be at Martinsburg, and intercept such as may attempt to escape from Harpers Ferry.

4. General Longstreet's command will pursue the same roads as far as Boonsborough, where it will halt, with reserve, supply and baggage trains of the army.

5. General McLaws, with his own division and that of General R. H. Anderson, will follow General Longstreet. On reaching Middletown will take the route to Harpers Ferry, and by Friday morning possess himself of the Maryland Heights and endeavor to capture the enemy at Harpers Ferry and vicinity.
6. General Walker, with his division, after accomplishing the object in which he is now engaged, will cross the Potomac at Cheek's Ford, ascend its right bank to Lovettsville, take possession of Loudoun Heights, if practicable, by Friday morning, Key's Ford on his left, and the road between the end of the mountain and the Potomac on his right. He will, as far as practicable, cooperate with General McLaws and Jackson, and intercept retreat of the enemy.
7. General D. H. Hill's Division will form the rear of the army, pursuing the road taken by the main body. The reserve artillery, ordnance, and supply trains, and c., will precede General Hill.
8. General Stuart will detach a squadron of cavalry to accompany the command of Generals Longstreet, Jackson and McLaws, and, with the main body of the cavalry, will cover the route of the army, bringing up all stragglers that may have been left behind.
9. The commands of Generals Jackson, McLaws and Walker after accomplishing the objects for which they have been detached, will join the main body of the army at Boonsborough or Hagerstown.
10. Each regiment on the march will habitually carry its axes in the regimental ordnance-wagons, for use of the men at their encampments, to procure wood and c.[7]

A careful reading of Special Order 191 divulges that it did not even require Jackson to go to Harpers Ferry. Furthermore, it tasked McLaws's two divisions with capturing Harpers Ferry, an impossible task as McLaws had only 8,000 men while the garrison had 14,000. Had the Confederate generals chosen to follow the letter of the law, the war in the east might have ended in September 1862. Fortunately for the Confederacy, they did not.

We have seen General McClellan's rather complex order to General Franklin on the 13th that resulted in the failure to relieve the Harpers Ferry garrison. Now let us look at Confederate General Jackson's order of the 14th that resulted in the successful capture of the garrison. The order was as follows:

Headquarters Valley District
Sept 14, 1862

I. To-day, Major General McLaws will attack so as to sweep with his artillery the ground occupied by the enemy, take his batteries in reverse, and otherwise operate against him, as circumstances may justify.
II. Brigadier General Walker will take in reverse the battery on the turnpike, and also sweep with artillery the ground occupied by the enemy, and silence his batteries on the island in the Shenandoah should he find a battery there.
III. Major General A. P Hill will move along the left bank of the Shenandoah, and thus turn the enemy's left flank and enter Harpers Ferry.
IV. Brigadier General Lawton will move along the turnpike for the purpose of supporting General Hill and otherwise operating against the enemy on the left of General Hill.
V. Brigadier General Jones will, with one of his brigades and a battery of artillery, make a demonstration against the enemy's right; the remaining part of his division will constitute the reserve and move along the turnpike.

By order of Major General Jackson[8]

It will be seen from the above that the actual capture of Harpers Ferry was entrusted to Major General A. P. Hill, and his instructions consisted of a single sentence.

7

Perryville

Our fateful battlefield order of this chapter relates to the battle of Perryville. However, to understand what happened, we must again start at the beginning and describe the events leading up to Perryville.

At the outset of the Civil War, the South unwisely tried to protect the entire border but lacked the resources to do so. The first cracks in its borders in the west occurred in the spring of 1862. It was a result of the Union navy's superiority not only on the high seas but on the inland waters as well. The entry points into the interior of Tennessee were the Cumberland and Tennessee Rivers, as well as the Mississippi. Unless the South could block entry into their waterways, the Union could proceed into the interior of the state practically unmolested. Island 10 blocked access to the Mississippi, Fort Donelson to the Cumberland, and Fort Henry to the Tennessee. In the spring of 1862, the South lost all three and the interior of Tennessee was now wide open.

The Confederate commander in the west, Albert Sydney Johnston, desperately tried to recoup the situation. His plan was to use the quickest means possible to gather all his widespread forces into a single army and then defeat the Union forces in detail before they could concentrate. The quickest point of concentration for the Confederate forces was Corinth, Mississippi, the rail junction of the Memphis-Charleston and Mobile-Ohio railroads.

The Confederates united at Corinth and on April 6, 1862, struck the nearby Union army of General Grant at Shiloh. They came within a hair's breadth of victory but then, on April 7, the Union Army of the Ohio, which had been on forced marches to join Grant, reached the scene and turned the tide. It was a close call and General Johnston, the Confederate commander, was killed in the battle.

The combined Union armies, now under General Halleck, slowly pressed forward and seized Corinth on May 29, 1862. It is at this point that our story of Perryville really begins.

The combined Union armies at Corinth now faced the combined Confederate armies at Tupelo. This situation was to be short-lived. In July, General Halleck ordered General Buell and his Army of the Ohio to depart the Corinth area and to resume its primary mission, the conquest of the remainder of Tennessee. General Bragg, now the commander of the combined Confederate forces, responding in kind, transferred the bulk of his army by a circuitous rail route to Chattanooga in east Tennessee (see map 18).

So by the end of July 1862, we had Buell's army of 40-some thousand moving eastward in southern Tennessee toward Chattanooga, and Bragg's army of 32,000 ensconced in Chattanooga, presumably beginning a contest for the control of east Tennessee.

However, unknown to the Union generals, the Confederate plans were far more ambitious than maintaining control of east Tennessee. They planned nothing less than the invasion of neutral Kentucky and adding it to the Confederacy as an 11th state.

If Kentucky were added to the Confederacy, the boundary of the nation would be pushed northward to the Ohio River. The Confederates would add immense industrial, agricultural, and personnel resources to their cause, and the Union position in Tennessee would be untenable.

The campaign for Kentucky would begin on August 16, 1862, with the two main contending armies, those of Buell and Bragg, hundreds of miles from the state.

Before we commence a narrative of the campaign, let us take a closer look at the leading players.

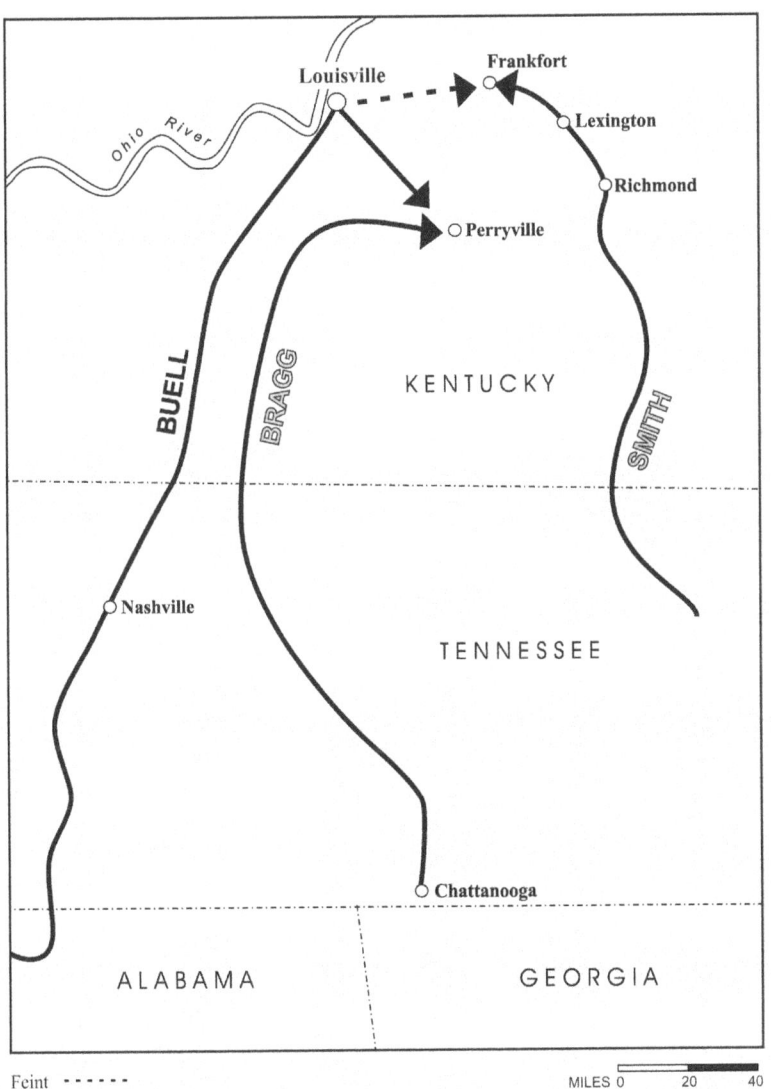

Map 18
Perryville

The Cast of Characters

Don Carlos Buell was the commanding general of the Army of the Ohio and the chief Union protagonist in the campaign for Kentucky and the battle of Perryville. At the time of the battle, he was 43 years old. He had graduated from West Point near the middle of his class in 1841, and participated in the Seminole War and the Mexican War. In the Mexican War, he was brevetted three times for bravery. At the outset of the Civil War, he was a brevet lieutenant colonel serving as the adjutant of the Department of the Pacific.

On May 17, 1861, he was jumped up in rank to brigadier general in the U.S. Volunteer Army. In November 1861, he was appointed head of the Department of the Ohio, and when he took to the field, commanding general of the Army of the Ohio. Thus, a man who had never before commanded more than 100 found himself in command of about 50,000 within six months. Buell, like several others at this stage of the war, owed his rapid advancement to the confidence of his friend, George B. McClellan, general-in-chief of the armies.

Buell was instrumental in organizing and honing the fine Army of the Ohio, capturing Nashville, and rescuing Grant at Shiloh. There is little doubt that he was a talented organizer and administrator, an unusually intelligent man, and a dedicated patriot. However, his manner was variously described as "cold" and "aloof."[1] This lack of charisma, or so-called common touch, precluded Buell from ever being loved or even popular with his troops in the manner of a Lee or a McClellan.

Another characteristic of Buell that ultimately proved his downfall was that he preferred to make war "light" like his friend McClellan. He lacked the killer instinct of a Patton or Jackson. To him, war was more like a chess match than a meat grinder. Buell himself stated:

> The object is not to fight great battles, and storm impregnable fortifications, but by demonstration and maneuvering to prevent the enemy from concentrating his scattered forces ... the commander merits condemnation who, from ambition or ignorance, or a weak submission to the dictation of popular clamor and without necessary profit, has squandered the lives of his soldiers.[2]

Major General Don Carlos Buell USV: Preferred to make war "light."

7. Perryville

General Braxton Bragg CSA: A difficult senior.

Braxton Bragg has at times been referred to as the worst general in the Civil War. If not the worst, he usually at least makes the list of contenders. Bragg was born on March 22, 1817, and thus at the outset of the Civil War was one year older than Buell.

Bragg graduated from West Point in 1837, fifth in a class of 50. He participated in both the Seminole War and the Mexican War, during which he received brevets for bravery and distinguished service. Bragg remained in the army until 1856, when he resigned to become a sugar planter in Louisiana. In this capacity, he became close friends with future Union general William T. Sherman, who for a time was president of Louisiana State University.

At the outset of the Civil War, Bragg was a major general in the Louisiana State Militia, which was soon to be incorporated into the Confederate army. By this time, Bragg had already acquired a reputation as a strict disciplinarian with an acerbic and confrontational personality.

After a number of unexciting commands, Bragg was assigned as a corps commander in Albert Sydney Johnston's army shortly before the battle of Shiloh. As luck would have it, or perhaps fate, Johnston was killed in the battle and shortly afterwards, his replacement, Beauregard, had to relinquish the command to go on sick leave. This left Bragg the senior officer on the scene. Bragg was then promoted to the rank of full general, an exalted position that only six other Confederates achieved. Bragg then assumed the command of the armies at Tupelo, and it is here that our story begins.

Bragg was never popular with the rank and file and quickly alienated the officers closest in rank to him. The most basic law in warfare is to feed victory and starve defeat. Bragg apparently never assimilated this correctly and always operated under the law "starve victory."

The discontent of Bragg's immediate subordinates was usually limited to back door political attempts to have him relieved. However, later in the war, one of his generals, Nathan Bedford Forrest, probably the most talented of all, expressed to Bragg's face what the others probably merely thought. Forrest said to Bragg in the presence of others, "You have played the part of a damn scoundrel.... If you ever try to interfere with me or cross my path it will be at the peril of your life."[3]

Our next character in this drama was Bragg's leading subordinate, Leonidas Polk. Polk is probably the most interesting of all of our main players. He was both bishop and general. If other Union and Confederate generals had been jumped up from company commander to army commander, Polk was jumped to major general and military department head from a clergyman who had never commanded anything at all.

At 56, Polk was older than the others. He was born in 1806 and attended West Point, class of 1827, where he graduated eighth in a class of 38. Other cadets at the academy at the time included Jefferson Davis and Robert E. Lee, both of whom became life long friends of Polk. Later Lee wrote that Polk was considered "as a model for all that was soldierly, gentlemanly and honorable."[4]

In his last year at the Point, Polk was drawn to the calling of the cloth. Soon after graduation, he resigned from the army and entered the Virginia Theological Seminary. Polk was ordained a deacon at 24, priest at 30, and bishop at 32. At the outbreak of the Civil War, he was the senior bishop of the Episcopal church in the south.

For reasons unknown, Polk's friend Jefferson Davis offered Polk a commission as major general. Polk accepted but did not lay aside his robes as bishop. He still presided at times over religious services.

Polk was an affable, likeable man and was always wildly popular with the rank and file, who delighted in stories of his eccentricities. One probably apocryphal story had Polk addressing his troops as they were about to go into action. General Cheatham said, "Boys, give them hell." Polk then chimed in, "Boys, give them what General Cheatham said."

Aside from Polk's attractive qualities, he had a shortcoming. He valued his own opinion very highly and liked to be always right. He had the disturbing habit of considering an order he did not agree with as a suggestion to be implemented or not as he saw fit. This ultimately brought him into conflict with his very demanding boss, General Bragg. At the time of our narrative, Polk was senior corps commander in Bragg's army and second in command to Bragg.

The last of our four main players is Confederate general Edmund Kirby Smith. Smith was not a subordinate of General Bragg, but a commanding general of a separate small army that was to cooperate with Bragg in the seizure of Kentucky.

Smith was the youngest of the four, being only 38 at the time of the Kentucky campaign. He, like the others, was a West Point graduate, graduating 25th out of 41 in the class of 1845. A classmate of his was Ulysses Grant. Smith participated in the Mexican War, but did not achieve the distinction of Bragg or Buell.

At the outset of the Civil War, he was given a colonel's commission and participated in the first battle of Bull Run. Here, the newspapers incorrectly portrayed him as a hero, which nevertheless furthered his advancement.

Luck seemed to be a hallmark of Smith's career. He always happened to be in the right place at the right time, and while never really accomplishing anything outstanding, ultimately became one of seven Confederates to achieve the rank of full general. He was the last of the full generals to surrender and the last to die.

The Invasion

The plan for the seizure of Kentucky by the Confederates actually entailed two separate armies. The first and largest, of course, was that of General Bragg. The second and much smaller army was initially called the Army of East Tennessee, but upon the invasion, was rechristened the Army of Kentucky. It was commanded by General Edmund Kirby Smith. Bragg's army contained about 32,000 men and Smith's 19,000. Unfortunately for the campaign, neither was in charge of the whole campaign. Each reported separately to Richmond. This did not bode well for the future as there is an old saying that "one bad general is better than two good generals."

The operation commenced on August 16, 1862, when Smith's small army entered Kentucky from the nearby Knoxville, Tennessee, area. At the time, Bragg's army was far away in Chattanooga and thus in no position to imminently join Smith or to act in close concert with Smith's movements.

The opposing army of Buell, consisting of some 40,000 men, was also far from Kentucky at the outset. A mobile force of about 25,000 men was in the vicinity of Huntsville, Alabama, and the other 16,000 were devoted to protecting Buell's long and vulnerable supply line that extended through Nashville, Tennessee, all the way to Louisville, Kentucky, a distance of 250 miles. Concurrent with Smith's entry into Kentucky, the armies of Bragg and Buell began their long marches to Kentucky.

Although the two armies followed roughly parallel routes and at times came within a few miles of each other, they never clashed before reaching Kentucky. Buell finally reached Louisville on September 24. Bragg just might have reached Louisville first and cut Buell off from his supplies, had he not digressed to surround and capture the Union supply base at Munfordville. This netted Bragg 400 Union prisoners.

While the armies of Bragg and Buell were trudging northward toward Kentucky, the army of Smith was rampaging through the state almost unopposed. A scratch force of 6,500 Union troops attempted to stop Smith before Lexington and Frankfort but, on August 29 and 30, in the battle of Richmond, the Union force was utterly destroyed in a one-sided battle. Smith went on to occupy both Lexington and the capital, Frankfort. Thus, by the time Buell's army finally arrived at Louisville, the bulk of Kentucky, including the capital, was in Confederate hands.

By October 1, we were rapidly approaching a showdown for the battle of Kentucky. All three armies were now in the state and within little more than a day's march from each other. Buell was at Louisville, Smith 45 miles to the east at Frankfort, and Bragg, completing the triangle about 40 miles southeast of Buell and a similar distance southwest of Smith.

By October 1, the balance of power was shifting. The Confederates had hoped to augment their armies with large numbers of Kentucky volunteers. In this they were disappointed. Few joined, in fact, not even enough to cover their attrition. However, while the Confederate armies waned, that of Buell waxed. En route to Louisville, he was augmented by two divisions sent from the army of Grant, and in Louisville he picked up an additional large contingent of Union troops that had been gathered there. By October 1, his army significantly exceeded the combined Confederate armies and he was ready to take the offensive.

As of October 1, although the ultimate outcome was still in doubt, the Confederates, at least for the moment, did control most of the state, did occupy the capital, and did have in their possession a Confederate sympathizing Kentuckian who had a legal claim, however tenuous, on the governorship. This was a certain Richard Hawes. These would legitimize the entry of Kentucky into the Confederacy. Then, the laws of the Confederacy would prevail. They could then accomplish by conscription what they were unable to accomplish by recruiting. They would install Hawes as governor in an inauguration ceremony at the capitol on October 4. They would even have an inaugural ball.

At the beginning of October, Bragg left his army, which was spread out between the towns of Perryville and Harrodsville, for Frankfort to attend the inauguration ceremonies.

At this point, we must go back a bit. In early October 1862, an almost unprecedented drought was raging in northern Kentucky; wells and streams were drying up, and even rivers were reduced to isolated puddles.

An army at the time must have water to move, and gobs of it. A typical army corps had 3 to 5,000 horses and mules. The officers had horses, the staffs had horses, the couriers had horses, the attached cavalry contingent had horses, the artillery was pulled by horses, and the entire transport system was horse drawn. Each horse drank several gallons a day, so the water requirement was not just for men filling their canteens at some farmer's well. The requirement was for tens of thousands of gallons a day.

Buell looked over the situation. He had two targets: Bragg's army at Perryville-Harrodsville, and Smith's army at Frankfort. Perryville had water. The main target would be Perryville. He would allocate two divisions to harass Smith at Frankfort while he would conduct the main attack using three corps against Perryville. Perryville was at a road junction with three roads converging from northwest to southwest, and meeting at Perryville. One corps would approach on each road. They would not attack until the three converging groups united and formed a single line of battle. The date of the attack would be October 8, 1862. In the preceding days, the two divisions would be pressing Smith at Frankfort, leading him to suspect that the main attack would be against him.

Smith and Bragg were at Frankfort for the ceremonies. General Hardee was in command of Bragg's troops at Perryville, and General Polk in command of those at nearby Harrodsville and also in overall command of the two.

On October 4, the inauguration ceremonies were proceeding swimmingly. The governor was sworn in. The ball was about to begin. The music started. But then ... then ... then gunfire was heard in the distance. The ball had to be postponed. As of now, it has been postponed over 150 years and the girls are getting old.

By October 6, the Union troops were pressing Smith so hard that he thought the main Union attack was going to be directed at him, and he asked Bragg for reinforcements.

Meanwhile, back at Perryville, the three Union corps were proceeding toward Perryville on three separate converging roads. Their last bivouac before the planned attack on the 8th would be the night of the 7th. The southernmost corps arriving in its camp area determined that there was no water. They heard, however, that water could be obtained two and a half miles off to the side, and thus continued on. This, of course, would widen the time that they would ultimately converge with the other corps on the 8th.

Bragg had an excellent cavalry commander in the person of General Joseph Wheeler. Wheeler thoroughly reconnoitered the approaching three Union corps throughout the 7th. He determined that the middle corps would arrive in the vicinity of Perryville much earlier than the other two. The head of its column should begin arriving around dawn; the northern column in the early afternoon; and the southern column, which had moved off its route for water, still later. The obvious course for the defending Confederates then was to attack the middle column at daybreak before either of the others could arrive. Wheeler so advised

General Hardee, the Confederate senior then at Perryville. Let us turn to Wheeler's own words:

> After dark at General Hardee's request, I went to his bivouac and discussed the plans for the following day. I explained the topography of the country and the location of Buell's columns. I understood from him that the attack would be made very early the next morning, and I endeavored to impress upon him the great advantage which must follow an early commencement of the action. An early attack on the 8th would have met only the advance of Gilbert's corps on the Springfield Road, which was four or five miles nearer to Perryville than any other Federal troops, and their overthrow could have been accomplished with little loss, while every hour of delay was bringing the rear divisions of the enemy nearer to the front, besides bringing the corps of McCook and Crittenden upon the field.[5]

Bragg, still not with his army, looked over the intelligence available to him and concluded that the force before Perryville was small. He thus ordered Polk to attack immediately. Bragg had thus come to the right actions for the wrong reason. Bragg's message to Polk was as follows:

> Headquarters Department No. 2
> Harrodsburg KY
> Oct. 7, 1862 — 5:40 P.M.
> General Polk:
> GENERAL: In view of the news from Hardee you had better move with Cheatham's division to his support and give the enemy battle immediately; rout him, and then move to our support at Versailles. Smith moves forward to-day in that direction, and I wish Withers to march to-night, toward Lawrenceburg, crossing thence tomorrow to Versailles, and follow up Smith and report to him. His wagon, except the ammunition and ordnance, had better cross at McCown's, turning off at Salvisa. No time should be lost in these movements. I shall follow Smith.
> Respectfully and truly, yours
> Braxton Bragg, Commanding[6]

This order appears clear enough as regards to what Polk was expected to do. He was to move with Cheatham's division to Hardee's support (at Perryville) and give battle immediately. The order even ended with the statement, "No time should be lost in these movements." This was exactly the right thing to do. It was what Wheeler had recommended to Hardee that date. The Confederates were to attack the first of the three Union corps arriving at Perryville before the other two could come to its support.

Inasmuch as Bragg's message was timed 5:40 P.M. October 7, it would be delivered to Polk that night. Any reasonable person would consider that the order to attack "immediately" would mean at first daylight on the 8th.

After sending his order, Bragg himself set out for Perryville. He arrived at 10 A.M. on the 8th. He had every expectation of either seeing a battle in progress or a victory celebration for a battle won. When he arrived he found that, not only had nothing happened, but even that nothing was in progress or preparation. He was furious and ordered that the attack be made immediately.

It was 2 P.M. before the attack was fully delivered. By that time, the second Union corps was already arriving on the field. Despite this, the attack appeared to be successful and drove all before it. However, late in the afternoon, the third Union corps arrived, and the Confederates had advanced so far that it hit them in the flank. This and darkness brought the battle of Perryville to an end. The score: Union casualties — 4211; Confederate casualties — 3396.[7]

Before we leave the battle of Perryville, let us recount an incident in the battle involving Polk as described by Polk to a foreign observer after the battle.

> "Well, sir," said Polk, "it was at the battle of Perryville, late in the evening — in fact, it was almost dark when Liddel's brigade came into action. Shortly after its arrival, I observed a body of men, whom I believed to be Confederates, standing at an angle to this brigade, and firing obliquely at the newly arrived troops. I said, 'Dear me, this is very sad, and must be stopped,' so I turned round, but could find none of my young men, who were absent on different messages; so I determined to ride myself and settle the matter. Having cantered up to the colonel of the regiment, which was firing, I asked him, in angry tones, what he meant by shooting at his own friends, and I desired him to cease doing so at once. He answered with surprise, 'I don't think there can be any mistake; I am sure they are the enemy.'
>
> "'Enemy!' I said; 'why I have only just left them myself. Cease firing sir: what is your name, sir?'
>
> "'My name is Colonel ____, of the ____ Indiana; and pray sir, who are you?'
>
> "Then, for the first time I saw, to my astonishment, that he was a Yankee, and I was in rear of a regiment of Yankees. Well; I saw there was no hope but to brazen it out. My dark blouse, and the increasing obscurity befriended me; so I approached quite close to him, and shook my fist in his face, saying, 'I'll show you who I am sir! Cease firing, sir, a once.' I then turned my horse, and cantered slowly down the line, shouting in an authoritative manner to the Yankees to cease firing; at the same time I experienced a disagreeable sensation, like screwing up my back, and calculating how many bullets would be between my shoulders every moment. I was afraid to increase my pace until I got to a small copse, when I put the spurs in and galloped back to my men. I immediately went up to the nearest colonel and said to him 'Colonel I have reconnoitered those fellows pretty closely and I find no mistake as to who they are; you may get up and go at them.'"[8]

Polk's failure to obey Bragg's order of October 7 to attack immediately actually constituted the second instance of Polk failing to comply with an order of Bragg within one week. On October 2, while Bragg was with Smith at Frankfort for the inauguration ceremonies, the divisions of Buell sent to attack Frankfort were detected. Bragg, in consequence, set the following order to Polk:

> Lexington KY, October 2, 1862 — 1 P.M.
> Maj. Gen. Leonidas Polk
> Commanding Army of the Mississippi
> Bardstown KY:
>
> GENERAL: The enemy is certainly advancing on Frankfort. Put your whole available force in motion by Bloomfield and strike him in flank and rear. If we can combine our movements he is certainly lost. Your information of the 30th was correct, but your courier was two days and nights getting here. Dispatch me frequently to Frankfort.
> Yours Truly
> Braxton Bragg
> General Commanding[9]

This, like the order of October 7, was a direct, nonconditional order. Polk did not comply, but did send a reply to Bragg dated October 3, 1862, 3 P.M., stating why he considered it "inexpedient" and "impracticable" to comply and that he was going to "pursue a different course."[10]

We do not know what the ramifications of Polk's failure to comply with Bragg's order of October 2 are. However, it is easy to assess the damage to the Confederate cause by Polk's failure to comply with Bragg's order of October 7. The battle of Perryville proved to be the deciding battle for Kentucky. After Perryville, Bragg conceded the state to Buell and returned to Tennessee. But for Polk's failure to execute Bragg's order, Bragg just might have won Perryville decisively.

The Confederates attacked at Perryville at 2 P.M. By that time, a second Union corps was already deploying on the field and a third still several hours behind. Even under these circumstances the Confederates came close to success. Had they attacked at dawn as the order required, or as General Wheeler recommended, they would have had more than ample time to roll up the first corps before the second arrived.

Why then did Polk not comply with Bragg's order of October 7? Here we will turn to Polk's own words from a letter to General Hardee:

> As to the Perryville affair, if I am to be tried for the disobedience of orders there the question arises, what orders. Surely not what purports to be orders in the paper sent you and by you to me. That paper is not mandatory, but simply suggestive and advisory:
>
> "In view of the news from Hardee you had better move with Cheatham's division to his support and give battle immediately, etc. ... No time should be lost in these movements."
>
> The order was not "You will move upon Perryville and attack the enemy early next morning," as the paper sent you charges. The writing sent me was not an order at all, but counsel or advice to do a certain thing in view of information received from Hardee. It does not help the matter to say that I was advised to do it immediately, and that it was added that no time should be lost in profiting by the advice to rout him, etc. The language was clearly not peremptory, but suggestive and advisory, and left me the use of my discretion, as to the details of the attack, it being understood that I accepted the advice and proceeded to carry the operations into execution as judiciously and promptly as a willing mind etc., etc., etc., etc., etc.[11]

One might characterize Polk's defense as a mixture of sophistry and nonsense. Apparently, he considered that, had the order been worded "You will move upon Perryville and attack the enemy early next morning," it would have constituted a nondiscretionary order. Thus, the fate of a battle, a campaign, and perhaps the war hung on the wording of a single sentence. Perryville proved to be the decisive battle of the Kentucky campaign. Afterwards, Bragg's and Smith's armies vacated the state and conceded it to the Union. The Confederates had come close to winning and winning big at Perryville. Had they attacked at dawn, as Bragg desired, rather than at 2 P.M., their chances for a major victory would have been vastly enhanced. Had they won a major victory, they might well have won the campaign. Had they won the campaign and added Kentucky to the Confederacy, they might have won foreign recognition. Had they won foreign recognition, they might have won their independence. It all came down to words.

8

Fredericksburg

The Leading Players

The two leading Union players in the battle of Fredericksburg were Generals Ambrose Burnside and William Buel Franklin.

Major General Burnside is yet today one of the most interesting figures of the war. The prevailing view is that he was the incompetent par excellence. Others, however, contend that he was merely a victim of bad breaks, and actually a man of some considerable talent. Lincoln, a man who was a good judge of character, saw capabilities within him.

Burnside graduated from West Point in 1847 near the bottom of his class. He was too late for the Mexican War and spent five years of undistinguished service. He resigned in 1852 to pursue the manufacture and marketing of a breech-loading carbine that he had invented. The effort was a failure and he went into bankruptcy. He then turned to his friend, George McClellan, a railroad executive, for help and was given employment. The long friendship of Burnside and McClellan was not that of two equals. Rather, Burnside always looked up to McClellan as his superior.

At the outset of the war, Burnside received a commission in the rapidly expanding Volunteer Army as a colonel. He was quickly upped to brigadier general. In the fall of 1861, Brigadier General Burnside was given an important assignment through the intervention of McClellan. Burnside was designated commander of the so-called Coast Division. This was actually a combined land-sea force that was to conduct operations against Confederate enclaves on the North Carolina coast. The expedition stood out as a success at a time of a sea of Union failures. It conquered Roanoke Island on February 7–8, 1862; Elizabeth City on February 10; New Bern on March 14; and Fort Macon on April 26. These successes brought Burnside to the attention of the administration and, on March 18, 1862, he was promoted to major general.

Burnside played no part in the second battle of Bull Run and, as we have seen in chapter 6, his performance in the Maryland Campaign was pedestrian at best. Be that as it may, by the end of the Antietam campaign, Lincoln was totally dissatisfied with McClellan as army commander and was looking for a replacement. Burnside was next in seniority and resting on his laurels from the North Carolina coast campaign. Lincoln offered him command of the army. Burnside, being a modest man, did not feel that he was qualified and, in any event, did not want to offend the man he idolized — McClellan.

To get Burnside's true feelings, let us turn to his own words as expressed before a joint committee of Congress:

> ... that I did not want the command; that it had been offered to me twice before, and that I did not feel that I could take it ... I told them what my views were with reference to my ability to exercise such a command, which views were those I had always unreservedly expressed — that I was not competent to command such a large army as this. I had said the same over and over again to the President and Secretary of War; also, that if things could be satisfactorily arranged with General McClellan, I thought that he could command the Army of the Potomac, better than any general in it.[1]

The end came for McClellan late on the night of November 7, 1862. Amazingly, the message was handed to McClellan by the courier while McClellan was in his tent and involved in an amicable conversation with Burnside. The message was as follows:

General Orders, No. 182
War Department, Adjutant-General's Office
Washington, November 5, 1862
   By direction of the President of the United States, it is ordered that Major General McClellan be relieved from the command of the Army of the Potomac, and that Major General Burnside take command of that army.
   By order of the Secretary of War
   E. D. Townsend
   Assistant Adjutant General[2]

Major General Burnside was now the man!

Our second main player was Major General William Buel Franklin. Franklin was a rarity in the upper echelons of both armies. While the top ranks of both armies were dominated by West Pointers, Franklin had finished at the top of his class. Even Lee and McClellan had failed to do this — both graduated second.

Franklin graduated in the class of 1843. Among his classmates was Ulysses S. Grant. Because of Franklin's exalted academic standing, he was naturally commissioned in the prestigious corps of topographical engineers. He spent the next 18 pre-war years in either engineering or administrative duties. He participated in the survey of the Rocky Mountains; hydrographic surveys; the building of lighthouses; and, as the secretary of the Lighthouse Board, controlling all

Major General Ambrose Burnside USV: Incompetent or misunderstood?

lighthouse construction. In 1859, he was assigned to what was probably the most prestigious job of all for army engineers. He was designated supervising engineer for the construction of the dome of the national capitol and the building's expansion. In this capacity, he dealt personally with political figures of the highest rank. Just before the outbreak of the war, Franklin was appointed supervising architect for the construction of the new treasury building. Franklin's name is thus immortalized by appearing on the name plaques of some of the government's most important buildings.

Despite Franklin's great success as an engineer, the outbreak of the war still found him a lowly captain. When wartime promotions were handed out, Franklin had one huge asset in addition to his undoubted mental acumen. He was a good friend of George B. McClellan. At the outset of the war, many young men who had never commanded more than a company of 100 men were quickly elevated to command brigades and divisions. Franklin, however, who had never commanded anything at all was, within one year, elevated to corps command. His sequence of promotions was as follows: May 14, 1861, promoted to colonel; May 17, 1861 (just three days later), promoted to brigadier general; May 8, 1862, assigned to command the newly formed Sixth Corps. This was still not all. When Burnside took command in December 1862, he gathered the corps into "grand divisions" of two corps each, and Franklin was placed in command of a grand division. Here now we had a man commanding at the very top who never had commanded at the bottom.

Franklin's corps was pushed front and center for the first time at the battle of South Mountain in September 1862. He seemingly scored a success but, actually, it was a colossal failure. He should have and could have relieved Harpers Ferry, but did not. Had he done so, the war in the east just might have come to a conclusion on September 17, 1862.

Now we come to Fredericksburg in December 1862. Franklin's command will again be pushed forward front and center. It is his command and his actions that will determine the outcome of the pending battle.

Major General William B. Franklin USV: A frequent central figure in "Bad Outcomes."

The Situation

At the time of Burnside's appointment, the Union Army faced the Confederate Army of Northern Virginia across the Rappahannock River near Rappahannock Station. This was on one of the two major north-south routes into the interior of Virginia and on to Richmond. The other route was 22 miles farther down the Rappahannock via Fredericksburg. The Washington-Fredericksburg route was the more direct route to Richmond.

Burnside was acutely aware of the

reason for McClellan's relief and his own appointment. He knew Lincoln wanted action and felt he must do something, and do it now. He successfully hoodwinked the Confederates and stole a march on them down the river to the Fredericksburg area. Fredericksburg was on the Confederate side of the river, and Burnside had sent a message to Washington requesting that pontoons for the building of bridges be awaiting him by the time his troops arrived opposite Fredericksburg. However, true to form, Washington failed him; and by the time the pontoons finally arrived, the Confederate Army was ensconcing itself on the other side of the river to contest any crossing. However, all was still not lost. The Rappahannock was still bridgeable for more than 25 miles beyond Fredericksburg before it widened and finally emptied into the Chesapeake Bay. Perhaps Burnside could still make an uncontested crossing further down. After all, he could pick any point, while the Confederates, not knowing the point, had to spread out to protect all points. As the two armies squared off on opposite sides of the Rappahannock, the Union Army present totaled about 120,000, while the Confederate Army was about 75,000.

To the surprise of Burnside's subordinates, as well as the Confederates, he decided to build his bridges directly opposite Fredericksburg—in the face of the main body of the enemy. The crossing would lack surprise and have to be forced.

But, to go back a bit, Burnside, upon taking command of the army, decided upon a major reorganization. The army consisted of seven corps plus cavalry. The corps commanders plus the cavalry commander reported directly to the army commander. Burnside wanted to simplify things and reduce the number of individuals reporting directly to him. Consequently, he created three "grand divisions," each consisting of two corps, and left the seventh, remaining corps as the "reserve corps." Now, he just had the three grand division commanders plus the reserve corps and cavalry commander reporting directly to him. The three grand division commanders were Generals Franklin, Hooker, and Sumner. It was these that were to fight the forthcoming battle of Fredericksburg. The reserve corps was not present, and the cavalry played little part.

By December 12, Burnside had succeeded in building five pontoon bridges over the Rappahannock, two directly opposite the city of Fredericksburg and three about three miles down river. The plan was for General Sumner with his grand division to cross the two bridges directly into Fredericksburg, General Franklin with his grand division to cross on the three bridges farther down, and General Hooker with his grand division to remain in reserve on the north side of the river. Hooker was directed to station two of his six divisions at the foot of the three bridges to support Franklin as required, and four of his six divisions at the foot of the two bridges to support Sumner as required.

General Lee offered some resistance to the building of the two bridges directly into the city, but decided not to make his main resistance at the river's edges but rather, at a ridge some one to two miles south of the river that ran roughly parallel to the river (see map 19). The ridge defense line was about seven miles long and covered all roads and the railroad leading south. The Confederates had constructed a military road just behind the ridge, allowing them to quickly move troops from one end to the other. Lee's army was divided into two corps plus cavalry. The corps were headed by Generals Longstreet and Jackson, and the cavalry by General Stuart. Longstreet covered the Confederate west end of the ridge, Jackson the east end, and Stuart the open flat space beyond the ridge to the east. Longstreet's segment of the ridge had great natural strength. His end of the ridge was

Map 19
Fredericksburg — The Battle of Fredericksburg

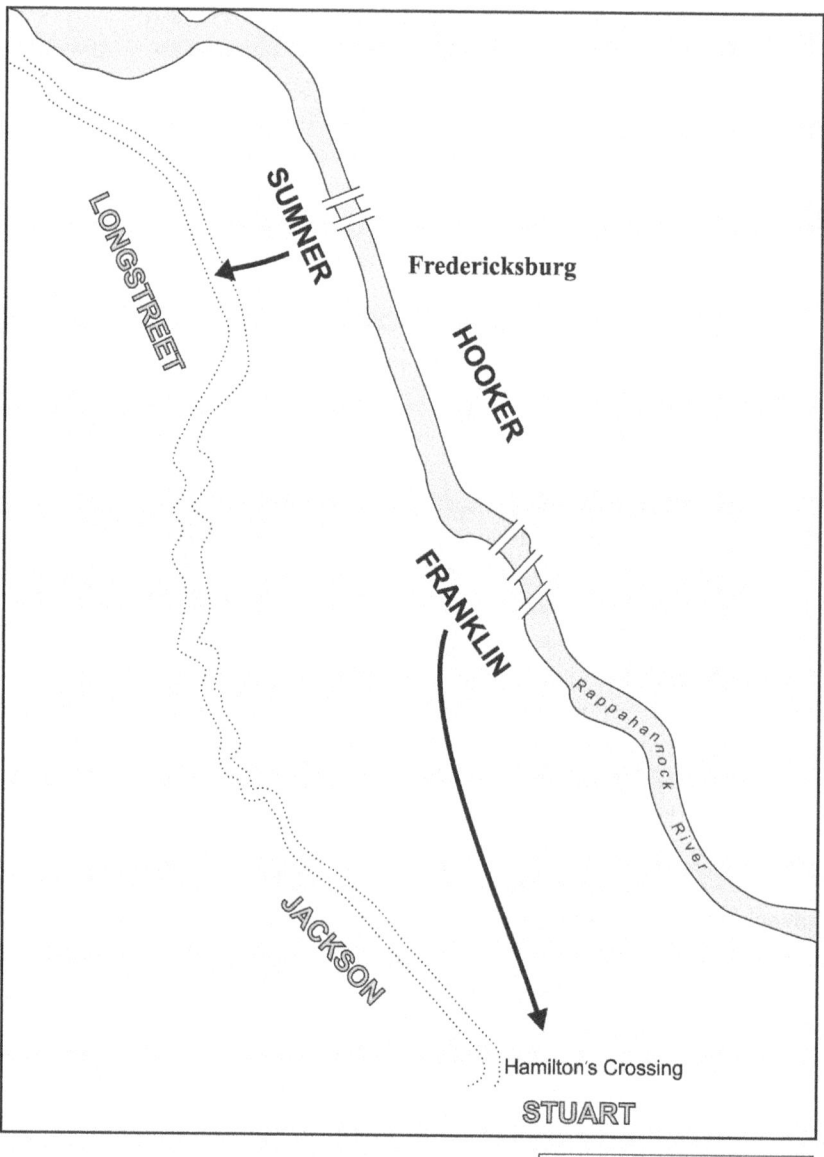

anchored on a canal and then on a bend of the Rappahannock. Consequently, it could not be flanked. For most of Longstreet's front, there was a sunken road along the base of the ridge on the Fredericksburg side that was bordered by a stone wall. The defending soldiers thus had an almost shoulder high bullet-proof protection. Infantry behind the wall, plus artillery on the ridge facing an open approach, presented an almost impregnable defense.

Jackson's section was by no means as favorable for defense. There was no sunken road or wall at the base of the ridge, and the ridge terminated at a place called Hamilton's Crossing. The name was derived from the fact that the railway from Fredericksburg to Richmond angled around the end of the ridge here in its journey south, and crossed the north-south highway. There was no natural obstacle to prevent the flanking of the ridge line at this point. It was obvious to anyone familiar with the terrain that, if the Confederates were to be pushed out of their ridge position, the key was here. If Hamilton's Crossing could be taken, the Confederate ridge line could be rolled up from east to west. Pressure here would inevitably draw defending troops from the strong western part of the line.

The bridges were complete by the 12th and Burnside ordered the crossing of the river — Sumner by the two bridges opposite Fredericksburg, and Franklin by the three further down. The crossings were successfully completed during the day, and Hooker, with his grand division in support, took up his positions — two divisions at the three bridges behind Franklin's position, and four divisions at the two bridges behind Sumner's position. By early afternoon on the 12th, Burnside was about ready to begin the main stage of the battle of Fredericksburg. The attack on the Confederate ridge position would take place on the morrow, December 13, 1862.

At this point, with his two grand divisions over the river and the third waiting at the bridges in support, Burnside rode the line for a final conferencing with his generals. He arrived at Franklin's headquarters at about 5:00 P.M.[3] Fortunately for history, Franklin had invited his two principal assistants, his corps commanders Generals Smith and Reynolds, to attend the conference. Subsequently, Franklin, Smith, and Reynolds all were to agree on what transpired. Franklin recommended to Burnside that the main attack should be performed by his grand division, that the initial objective should be the ridge just above Hamilton's Crossing. Franklin proposed using at least 30,000 of the 40,000 men of his command in the initial attack. He would form these men into two assault columns during the hours of darkness, and attack at daylight. However, in order to do this, Franklin pointed out that it would be necessary for the two divisions of Hooker's grand division that were waiting at the other side of the bridges to cross over during the night to relieve two of his divisions that were tasked with guarding the bridge approaches. The three generals agreed that Burnside appeared to assent to Franklin's plan. When Burnside finally departed at about 6:00 P.M., he indicated that he would order the two divisions of Hooker's grand division to cross over during the hours of darkness, and that in two to three hours time, Franklin could expect to receive a written order to confirm the operation.[4] As Burnside departed, Franklin felt that all was well.

Then Franklin waited ... and waited ... and waited. As it got dark and nothing happened, he became ever more apprehensive. Finally, at midnight, he sent Colonel McMahon to the nearby telegraph terminal to ascertain why the delay. He was assured that the order was under preparation and would be issued forthwith.[5] Midnight passed and no order. There was no movement on the bridge, and still no order. As the hours of darkness ticked away, Franklin became ever more frantic. To have the reserve divisions cross in the daylight and allow him to assemble his assault columns in full view of the Confederates would be suicidal. As dawn approached, Franklin, in desperation, sent a dispatch asking if there was anything for him.[6] No answer. Dawn broke and Franklin realized now that it was already too late to carry out his plan. Then the order arrived. It did not come over the wire or by courier. It came in the hands of a general from Burnside's staff. The general was Brigadier General Hardie, the assistant inspector general. The plan was handed to Franklin at 7:30 A.M. It had a time of origin of 5:55 A.M. The order read as follows:

HEADQUARTERS ARMY OF THE POTOMAC
December 13, 1862 — 5.55 A.M.
Major-General Franklin,
Commanding Left Grand Division, Army of the Potomac:

General Hardie will carry this dispatch to you, and remain with you during the day. The general commanding directs that you keep your whole command in position for a rapid movement down the old Richmond road, and you will send out at once a division at least to pass below Smithfield, to seize, if possible, the height near Captain Hamilton's, on this side of the Massaponax, taking care to keep it well supported and its line of retreat open. He has ordered another column of a division or more to be moved from General Sumner's command up the Plank road to its intersection with the Telegraph road, where they will divide, with a view to seizing the heights on both of these roads. Holding these two heights, with the heights near Captain Hamilton's, will, he hopes, compel the enemy to evacuate the whole ridge between these points. He makes these moves by columns distant from each other, with a view of avoiding the possibility of a collision of our own forces, which might occur in a general movement during a fog. Two of General Hooker's divisions are in your rear, at the bridges, and will remain there as supports. Copies of instructions given to Generals Sumner and Hooker will be forwarded to you by an orderly very soon. You will keep your whole command in readiness to move at once, as soon as the fog lifts. The watchword, which, if possible, should be given to every company, will be "Scott."

I have the honor to be, general, your obedient servant,
JNO. G. Parker
Chief of Staff[7]

As Franklin read on, he became ever more amazed. The points of attack were as agreed; i.e., Franklin's was Captain Hamilton's (Hamilton's Crossing) and Sumner's was the stone wall in Longstreet's sector. However, beyond this there were great differences. Instead of attacking with 30,000, Franklin was to "...send out at once at least a division" and Sumner's attack appeared to be similarly limited. Furthermore, instead of Hooker's two divisions being ordered to cross the bridge to support Franklin, they were ordered to "remain there as supports." Franklin finally decided that what Burnside wanted was merely a probe of the enemy strength at Hamilton's Crossing, to be followed later by a full attack to be ordered by Burnside.

Franklin ordered the small division of General Meade, which contained only 4,500 men, to execute the order to seize the heights at Hamilton's Crossing. The reason for selecting Meade was because, at the time, he was closest to Hamilton's Crossing. Franklin ordered the division of General Gibbons to protect the exposed right flank of Meade. This small force thus set out by 10:00 A.M. to do the seemingly impossible — seize the heights adjacent to Hamilton's Crossing.

Surprisingly, Meade, by chance, hit a weak spot in the Confederate lines and proceeded all the way to the heights. Being unsupported, however, he could not hold them, and by early afternoon was forced back to his starting point with great loss. Union losses in this limited attack were: Meade — 1,760, Gibbons — 1,249, Birney — 961.[8]

General Hardie faithfully submitted continuous reports by telegraph to Burnside on the status of Franklin's operations. Consequently, Burnside was fully informed of Meade's advance, repulse, and retreat. Franklin received no further instructions from Burnside of any kind until 2:25 P.M. on the afternoon of the 13th. The 2:25 P.M. order read, "Your instructions of this morning are so far modified as to require an advance upon the heights immediately in your front."[9]

This was strange. It could not help Meade. Meade had already been pushed back. Furthermore, "the heights immediately in your front" were certainly not the heights adjacent to Hamilton's Crossing. And Franklin stated later that immediately in the front of one of his divisions was a valley with rebels on the hills on both sides. Furthermore, he asserted it was already too late in the day, as it would be dark by five.[10]

Franklin was to receive his third and last order from Burnside for September 13 at 3:00 P.M. This order was delivered verbally by a staff officer. It was, as recalled by Franklin, that General Sumner's troops were being hard pressed, and could he make a diversion in Sumner's favor? This then was actually a request or suggestion rather than an order. With this, we conclude the sum total of orders that Franklin received from Burnside during the battle of Fredericksburg.

In the event, no "main attack" was ever made by Franklin on the 13th, and Sumner continued to waste his men on futile attacks against the unweakened Confederates behind the stone wall. The battle ended with the passing of daylight on December 13, 1862. The casualties were: Union — 12,653; Confederate — 5,309.[11] The Union army ended up back on the north side of the Rappahannock River as it began, and the casualties were for nothing.

The Aftermath — What Happened to Them?

In all, on that fateful day of December 13, 1862, 14 Union attacks were made against the stone wall. No Union soldier got even close to the wall. At dusk the ground was strewn with the dead and wounded. The weather was near freezing and many of the wounded would not survive the night. Lee, surveying the field, said, "It is well that war is so terrible! We should grow too fond of it!"[12]

The results provoked outrage in the northern press. The Union army had great superiority in men and equipment, and the rank and file had performed gallantly. Yet, they lost. It must have been the leadership. Someone had to be brought to account for the debacle.

As might be expected, Congress stepped into the picture. It was to do what Congress does best: talk, investigate, and assign the blame to others. The matter was taken under investigation by the Joint Committee for the Conduct of the War. Burnside and Franklin were called to testify. Burnside testified that his plan was as follows:

> The enemy had cut a road along the rear of the line of heights where we made our attack, by means of which they connected the two wings of their army, and avoided a long detour around through a bad county. I obtained from a colored man, from the other side of the town, information in regard to the new road, which proved to be correct. I wanted to obtain possession of that new road, and that was my reason for making an attack on the extreme left. I did not intend to make the attack on the right until that position had been taken, which I supposed would stagger the enemy, cutting their line in two; and then I proposed to make a direct attack on their front and drive them out of their works.[13]

The report concluded by blaming General Franklin for the defeat. It read:

> The testimony of all the witnesses before your committee proves most conclusively that, had the attack been made upon the left with all the force which General Franklin could have used for that purpose, the plan of General Burnside would have been completely successful, and our army would have achieved a most brilliant victory.[14]

The report said nothing about battlefield order writing which was the true culprit in the Union defeat.

Although Burnside seemed to have dodged the bullet in the congressional report that blamed Franklin, he did not do so in the eyes of the public or army. The officer corps and the rank and file both looked to him as the culprit. Morale plummeted and desertions increased.

Many of the senior officers, acting just short of mutinous, were now conspiring politically as to how to get rid of Burnside. Burnside knew that he must do something, and do something quick and dramatic, to restore his army's morale and reputation. Consequently, although it was already late in the season when he should be going into winter quarters, he planned one more quick campaign to turn the Union army's flanks. Within hours of the start, the rains came and the army became bogged down in the mud. Wagons and cannon simply could not be moved, and the infantry could not march. The whole thing had to be called off almost as soon as it started. Burnside had added a second fiasco to the first. Now there was "Burnside's Wall" and "Burnside's Mud March."

Burnside now considered that he would either have to get rid of the disaffected generals in his command or resign. He went to Lincoln and laid the choice before the president. Lincoln said he would consider the matter. He urged Burnside to stay, but finally accepted his resignation. This, however, was by no means the end of Burnside.

Burnside was offered command of the Department of the Ohio and accepted. This was not a humiliating reduction. The new command was important, and not far behind in importance to the command he had relinquished. Despite a couple of political faux pas, militarily Burnside did exceedingly well in his new command. He conquered Knoxville in east Tennessee and then successfully defended it against Longstreet's corps.

In early 1864, Burnside was placed in charge of the Ninth Corps and called east to assist Grant in his final campaign to seize Richmond and destroy Lee's army. Burnside participated in all the big battles leading to the final stage: the siege of Petersburg. It was here that Burnside met his Waterloo.

In the summer of 1864, mobile warfare for the Army of the Potomac came to an end when Lee's army dug in before Petersburg, a suburb of Richmond, for the defense of Richmond. Grant now laid siege to Petersburg/Richmond and Burnside's Ninth Corps occupied a section of the siege lines.

One of Burnside's regiments consisted of Pennsylvania coal miners. Their commander, Colonel Henry Pleasants, a mining engineer, approached Burnside with an idea. He said that his miners could dig a mine under the opposing Confederate entrenchments, pack it with gunpowder, blow a hole in the Confederate line, and march through to Richmond. Burnside thought it a good idea, secured the necessary permissions, and told Pleasants to go ahead.

The big day was July 30, 1864. The mine was dug, the powder packed, and the explosion set off. It worked to perfection. There was now a giant crater where the Confederate fortifications had been. What followed was chaos. Burnside's men hesitated and admired their work. They then marched into the crater instead of around it. The surviving Confederates gradually gathered around the rim and slaughtered Burnside's men. What had begun so auspiciously ended in tragedy. It was a Union defeat.

A court of inquiry was convened to assess blame. The blame was widespread but Burnside was singled out as the main culprit. He was relieved of command and told to await orders. None ever came.

Amazingly, he elected to go to Ford's Theater the night of April 15 to see *Our American Cousin*. He had no reason to believe that Lincoln would attend the same performance. Thus, fate would have it that he was present at Lincoln's assassination. Burnside resigned his commission the following day, April 16, the day Lincoln died.

Let us take a brief look at Burnside the man and Burnside the general. By all accounts, Burnside was an honest, modest man who readily admitted his faults and shortcomings. He was impressive in appearance. In a contemporary article in Harper's magazine, Burnside was described as "A very handsome man ... tall and stout, with a flashing eye and a sonorous voice, he looks the very beau ideal of a soldier."[15] His mannerisms were usually described as affable, genial, and similar terms.

As to Burnside the general, we have assessments by all three of Burnside's wartime commanders. Shortly after Antietam, McClellan wrote his secret assessment of Burnside in a letter to his wife. He wrote, "He is very slow, is not fit to command more than a regiment."[16] Grant in his memoirs wrote:

> General Burnside was an officer who was generally like and respected. He was not, however, fitted to command an army. No one knew this better than himself. He always admitted his blunders, and extenuated those officers under him beyond what they were entitled to. It was hardly his fault that he was ever assigned to a separate command.[17]

General Meade wrote the following about Burnside to his family shortly after the crater incident, "I feel sorry for Burnside, because I really believe the man half of the time doesn't know what he is about, and is hardly responsible for his acts."[18]

After leaving the army, Burnside went on to a brilliant civilian, political career. He was elected governor of his native state of Rhode Island three times and then served as one of Rhode Island's two U.S. senators until his death in 1881 at the age of 56.

Today, Burnside is chiefly remembered for a facial hair style that he made famous: originally "Burnsides," it somehow has come down to us as "sideburns."

Our other main player was Major General William Buel Franklin. Despite the fact that General Franklin was not really the man most responsible for the Fredericksburg debacle, he never entirely freed himself from the stigma of the congressional committee's accusation. Shortly after the battle, he was relieved from command and waited five months for new orders.

Franklin's new orders placed him in command of the Nineteenth Corps in Louisiana, in the domain of General Banks. In this capacity, he became a major participant in Banks's ill-fated Red River Campaign. The campaign came to an end when the Confederates defeated Banks's army in the battle of Mansfield on April 8, 1864. Franklin received a serious leg wound in the battle that effectively terminated his field service for the duration of the war. One could hardly blame Franklin for the failure of the campaign but, once again, he was a central figure in a major failure.

Franklin had commanded troops in numerous battles from Bull Run on. However, he came front and center to public attention in only three instances: South Mountain-Harpers Ferry, Fredericksburg, and the Red River Campaign. In all three instances, his name was associated with failure. Of the three instances, he was clearly at fault in only one, that of South Mountain-Harpers Ferry. Had he relieved Harpers Ferry, as he could have and should have, the war in the east just might have come to a conclusion on September 17, 1862.

Although Franklin's performance during the four years of the Civil War is often looked upon as a failure, he was a success in everything he undertook both before and after. Franklin resigned his commission in 1866 and entered civilian life. For the next 22 years, he was closely associated with the management of the Colt Firearms Manufacturing Company in Hartford, Connecticut, during which time he turned it around from near bankruptcy to great profitability. He also successfully held many midlevel civic positions and was largely responsible for the design and construction of the new state capitol building in Hartford.

For the last ten years of his life, Franklin served as the president of the Board of Managers of the National Home for Disabled Volunteer Soldiers, the predecessor of the Veterans Administration. Franklin died on March 8, 1903, at the age of 80.

In Conclusion

Let us see if we can recast the battlefield orders from Burnside as if they were written in accordance with the rules contained in the introduction.

December 12, 1862 — 8 P.M.
Copy to:
General Franklin
General Sumner
General Hooker

My intention is to drive the Confederates from their positions on and in front of the ridge behind the Rappahannock on December 13, and in the process inflicting maximum casualties. This will be accomplished by making the main attack against the Confederate right flank in the vicinity of Hamilton's Crossing. This will be accompanied by vigorous demonstrations on their other flank that will ultimately change into an attack. Assignments are as follows:
General Franklin:

You will undertake the main attack using all forces available to you. The attack will commence as soon as daylight and visibility conditions permit. Your initial objective will be the heights adjacent to Hamilton's Crossing. Once obtained, you will wheel and proceed down the ridge rolling up the Confederate line.
General Sumner:

Commencing as early as daylight and visibility conditions permit, you will conduct vigorous demonstrations before the Confederate lines in front of you. When you have determined that the Confederate position has been sufficiently weakened by their transfer of troops, to their other flank, to permit reasonable prospects of success, you will convert the demonstrations into a full scale attack using all the forces available to you.
General Hooker:

Immediately upon receipt of this order, you will order the two divisions of your Grand Division that are camped by the three bridges behind General Franklin, to cross the bridges and to report to General Franklin for further orders.

When requested by General Sumner, you will order the four divisions of your Grand Division that are camped by the two bridges behind General Sumner to cross the bridges and to report to General Sumner for further orders.
For all:

You are authorized to deviate from these orders and to use your own judgment in the following situations:

(A) If you encounter an unforeseen opportunity, the exploitation of which promises greater advantage

(B) If you encounter a situation where you consider that to proceed, you would be wasting the lives of your men without any realistic prospects of success

Keep me informed of your actions and intentions at a minimum by bi-hourly reports. I will be located at the telegraph terminal at the Phillips House in Falmouth.

General Burnside

Had Franklin received this order in place of those he did, would things have turned out more favorably for the Union?

As things actually turned out, Franklin never conducted a "main attack," never brought his biggest corps into action at all, while Sumner wasted his men in futile attacks on the unweakened Confederate left flank.

9

Vicksburg

The Mississippi River bisected the Confederate states. If the Union could get control of the whole river, the resource-rich west would be cut off from the more populous east, and the Union would be well on its way to final victory. As of April 1863, the Union controlled the north and the south of the Mississippi, but the Confederacy still clung to a stretch in the middle. The Confederate stretch extended 105 miles as the crow flies from Vicksburg, Mississippi, to Port Hudson, Louisiana.

This 105-mile stretch was vital to the Confederacy as the Red River from the west emptied into the Mississippi just above Port Hudson. Thus, the resources of the west could arrive by water to Vicksburg, which was on a railroad that extended all the way to the Atlantic Ocean (see map 20).

Both Vicksburg in the north and Port Hudson in the south had been converted into fortresses. Each was located on a high bluff towering above the Mississippi and each was located on a U-shaped bend in the Mississippi, with but a narrow tongue of land separating the prongs of the U. Thus, any passing ships had to pass the guns of the fortresses twice. In both instances, the guns were extensive, included naval guns throwing 100-pound shells, and were located both at water's edge and on the bluffs. Thus, any passing ships were subjected not only to direct fire, but to plunging fire from the bluffs, which was particularly effective against ironclads. In addition, if a passage were attempted at night, Confederates located on the tongues of the U's could light fires to silhouette the passing ships.

Vicksburg posed yet another impediment to its capture. The land to its north was all swamp, and in order to land troops, one had to negotiate many miles of swamp to reach dry land. Vicksburg was sometimes referred to as the Gibraltar of the Confederacy.

The Union forces to the north of Vicksburg were called the Army of the Tennessee and were commanded by General U.S. Grant. The Union forces to the south of Port Hudson were known as the Department of the Gulf and were commanded by General Banks. The naval forces collaborating with Grant were headed by Admiral Porter, and those collaborating with Banks were headed by Admiral Farragut. It is Grant's campaign to capture Vicksburg that is our subject here, so from this point, we will concentrate on it.

Grant had made several attempts to capture Vicksburg from the north during the fall and winter of 1862–63, but all had failed. By April 1863, he realized that if he were to succeed in its capture, he would have to approach it from the south, where there was dry land. He could march his army past Vicksburg on the west side of the river, but he would still have to get across. To do so, he would have to get a large number of ships past Vicks-

Map 20
Vicksburg — Confederate Control of the Mississippi

burg's batteries. He would need enough warships to provide absolute control of the river and enough transports to carry the troops, horses, artillery, and supplies across.

Up to this time, the Union navy forces both to the north and the south had demonstrated that the two Confederate fortresses were unable to stop all access to the Confederate zone. An occasional warship was getting through, and as of April 1863, seven were operating in the Confederate zone, and one of the seven was blocking the mouth of the Red River. However, this level of access was insufficient to give either the Union or the Confederates complete control of the waterway. Most Union warships entering the zone suffered some degree of damage, and there were no Union repair or resupply facilities in the zone. Last, but not least, once in, they had to run the batteries again to get out. If

Grant were to cross the river with his army, he needed both more warships and more transports.

In one sense, it was easier to pass Vicksburg down into the zone than to pass Port Hudson up into the zone. The Mississippi flows from north to south. Thus, ships passing down past Vicksburg added three knots to their speed, while those passing up past Port Hudson subtracted three knots from their speed. The ships passing downward were thus within range of the shore batteries a far shorter time.

Admiral Porter attempted a mass downward passing of Vicksburg the night of April 16. The flotilla consisted of seven ironclads, one ram, and three transports — a total of eleven ships. One transport was sunk, and the remainder of the ships suffered some damage but successfully completed the passage. A second mass passage of six transports was attempted the night of April 23. One was sunk, but the remaining five succeeded in getting through with minor damage.

The ships rendezvoused with Grant at Hard Times, on the west bank of the Mississippi, 25 miles south of Vicksburg (see map 21). Grant now had all the shipping resources he required to cross his army over to dry land on the east bank, but he had to hurry. Vicksburg was still intact and he could not rely on being resupplied by the river. He would have to conduct his campaign with the supplies he had with him. He began the crossing from Hard Times to Bruinsburg on the east bank on April 30.

Before we proceed with Grant's campaign, let us digress and take a brief look a the Confederate generals opposing him.

**Map 21
Vicksburg**

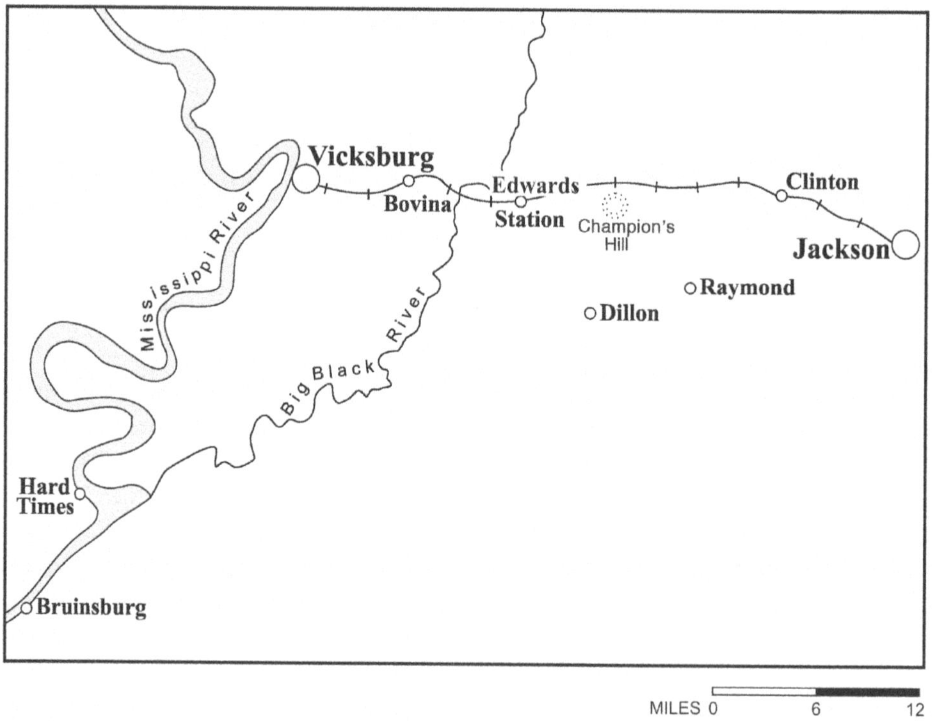

The Confederate Commanders

Lieutenant General J. C. Pemberton CSA: A yankee in rebel clothing.

Grant's opponent was Confederate Lieutenant General John Pemberton. Pemberton, like almost all of the top Confederate generals, was a West Point graduate. Pemberton graduated 27th of 50 in the class of 1837. His classmate and roommate was none other than the future Union General George Gordon Meade. In one of the amazing coincidences of history, roommates Meade and Pemberton were two commanding generals in the two greatest Union victories of the Civil War; Meade on the winning side at Gettysburg, and Pemberton on the losing side at Vicksburg. If that were not enough of a coincidence, the two victories occurred on the same day, July 4, 1863.

Pemberton remained on active duty from the time of his graduation until he entered the Confederate service in 1861. He had a modestly successful career in the old army. He won two brevets for gallantry in the Mexican War, but his regular rank at the time of his resignation to enter the Confederate services was just captain.

Pemberton was initially given a commission of lieutenant colonel in the Virginia state army, but was quickly promoted to brigadier general in the Confederate army. By October 1862, he had reached the exalted rank of lieutenant general, placing him near the very top of the Confederate hierarchy.

There is one thing that set Pemberton apart from all the other top Confederate generals. He was a Yankee from Pennsylvania who talked like a Yankee. He did not have a southern accent. A possible reason for his choosing the Confederate side was his wife, who was a Virginian and an ardent secessionist.

Pemberton, because of his background, never had the full trust of all of the leading southern politicians. However, he did have, and always retained, the trust and confidence of Jefferson Davis.

One of the characteristics of Pemberton, as we shall see, was indecisiveness. At the time of our narrative, he was commanding general of the Department of Mississippi and Eastern Louisiana. This encompassed both Vicksburg and Port Hudson, as well as most of the state of Mississippi, including its capital, Jackson.

We last left General Joseph Johnston in June 1862, when he was seriously wounded at the battle of Fair Oaks while serving as commanding general of the main Confederate army in the east. Johnston was relieved by Robert E. Lee. Johnston was one of the original five Confederate full generals, ranking fourth behind Samuel Cooper, Albert Sydney Johnston, and Robert E. Lee. He was small and slight of build, but of impeccable military bearing and always meticulously uniformed. He was always gracious with his subordinates and always extremely popular with the rank and file.

Johnston, however, was not well liked by his ultimate boss, Jefferson Davis. There were two bones of contention between Johnston and Davis. First, Johnston was the only one of the five Confederate full generals who had served in the rank of general in the old pre-war army. The other four had never risen higher than colonel. Johnston thus thought that it was he who should have been appointed the senior of the five. Second, Davis intensely disliked Johnston's propensity to continually retreat before the enemy. Johnston, unlike Davis and Lee, favored the Fabian approach to war. In his view, the Union had vastly superior manpower resources, and hence, he should only fight under those circumstances wherein he could inflict more casualties than he suffered. In short, while Davis and Lee thought that the South could win its independence on the battlefield, Johnston believed that they could only win by dragging out the war and making it so costly for the Union that the Union would gradually realize that it was not worth the trouble and allow the South to go.

On November 12, 1862, Johnston reported to the War Department that he was now fit to return to duty. He was assigned to what was to be a new command. The new command was to be headquartered in Chattanooga, and was to have jurisdiction over military operations between the Appalachians and the Mississippi. Thus, both Bragg's army in Tennessee and Pemberton's forces in Mississippi came under his jurisdiction. He had authority to transfer forces from one to the other or, if deemed necessary, to take command of either.

Thus, as Grant gathered his forces at Hard Times and prepared to cross the Mississippi, General Pemberton's immediate superior was Johnston.

Grant Crosses the Mississippi

On April 29, Grant had about 24,000 men at Hard Times, and by May 8, his strength in the area was to grow to about 34,000. Pemberton's strength at all times was about 40,000. But while Grant's forces were concentrated, Pemberton's were widely scattered. Pemberton had large forces at Vicksburg, Port Hudson, Jackson, and Grand Gulf and smaller forces scattered throughout his domain.

Grant started his crossing during daylight on April 30. By the end of the 30th, Pemberton knew that he was in big trou-

General Joseph E. Johnston CSA: Just too tolerant at Vicksburg.

ble. Grant had landed at Bruinsburg, 30 miles south of Vicksburg. From here, Grant was on solid ground and had multiple attractive options. He could move north and attack Vicksburg, he could move northeast and cut the railroad from Vicksburg to Jackson and isolate Vicksburg, or he could move on Jackson itself (see map 20). Jackson was not only the capital of one of the ten Confederate states, but a source of considerable production for the Confederate armed forces.

At least as late as May 8, Pemberton could have brought superior forces to the scene of action, denied Grant all three possibilities, and probably defeated Grant's force in the bargain.

The Confederate force in the immediate vicinity of Grant's landing was commanded by Brigadier General Bowen, and consisted of about 5,000 men. It could hope to do no more than delay Grant's advance.

Johnston was at Tullahoma, Tennessee, over 200 miles east of Vicksburg, on April 30 when he first learned of Grant's crossing of the Mississippi. Not only was he far away, but he was sick and considered himself unfit for duty. He was in communication with Pemberton by telegraph and dispatched his first order to Pemberton the following day, May 1. It read in part, "If General Grant's army lands on this side of the river, the safety of Mississippi depends on beating it. For that object, you should unite your whole force."[1] Johnston followed up with another message on May 2 that read in part, "If Grant's army crosses, unite all your forces to beat it. Success will give you back what was abandoned to win it."[2]

Pemberton, however, did not unite all his forces. He sent Bowen a meager 3,500 troops, which was only enough to ensure an orderly retreat back to Vicksburg. Pemberton, instead of uniting all his forces to put Grant on the defensive and seize the initiative, left the initiative with Grant and attempted to provide for all the contingencies that Grant might initiate — that is, protect Vicksburg, the railroad, and Jackson.

While many of the Confederate orders were put in the form of strong suggestions, Grant's orders to his subordinates were not. From the time of Grant's crossing of the Mississippi to the culmination of the campaign, Grant's control of his army might be likened to that of a conductor conducting a symphony orchestra. The imposition of his will on his subordinates was absolute. His orders were always crisp, clear, and unambiguous. A favorite phrase of his was "you will." In the absence of this, he sometimes simply ordered "move your command" or, occasionally, "I want you to..."[3]

His orders never left any doubt as to what the recipient was required to do. Additionally, they kept the recipient informed of what Grant was trying to do, what related commands were ordered to do, and what Grant thought the enemy was trying to do.

Grant's imposition of his will, when compared to Pemberton's tepid response to Johnston's orders, perhaps in part could be explained by the fact that Grant was present and might appear in person at any time, while Johnston was over 200 miles away.

Grant advanced toward the northeast and by the 13th had reached both Clinton and Raymond (see map 21). He was now on the railroad between Vicksburg and Jackson and within ten miles of Jackson on both the Clinton and Raymond roads. Pemberton's attempts to stop or slow down Grant had been ineffectual. He had failed to concentrate his forces, and Grant had triumphed over smaller Confederate forces at Port Gibson, Edwards Station, and Raymond.

Pemberton kept the War Department at Richmond and General Johnston at Tullahoma advised of these developments.

By May 9, the War Department became sufficiently alarmed that it was decided that something had to be done. That date, a telegram was sent to Johnston to proceed at once to Mississippi and to take command there. He was also directed to take 3,000 troops with him, and in the event they could not immediately proceed with him, he was to make arrangements for them to follow as soon as possible.

Johnston was still unwell, but replied by telegram, "Your dispatch of this morning received. I shall go immediately, although unfit for service."[4] The ailing, 56 year old Johnston took the first train available, which departed Tullahoma the morning of the 10th.

Johnston finally reached Jackson on the evening of the 13th, after a tedious three and a half day journey. Upon arriving, Johnston learned that the situation confronting him was as follows:

1. The Confederate brigades of Gregg and Walker, totaling about 6,000, were in Jackson.
2. Confederate General Maxey's brigade was expected to arrive in Jackson from Port Hudson on the 14th.
3. Reinforcements under General Gist were on the way to him by rail from the east, and should arrive on the 14th.
4. When Gist's reinforcements and Maxey's brigade arrived, he would have about 11,000 troops at Jackson.
5. Pemberton's forces, except two divisions in Vicksburg and 5,000 troops at Port Hudson, were at Edwards Station.
6. Four divisions of the enemy were already at Clinton on the Vicksburg-Jackson railroad between Pemberton at Edwards Station and himself in Jackson. Clinton was just 15 miles east of Pemberton at Edwards Station and ten miles west of Johnston at Jackson.

Johnston concluded that his force at Jackson must unite with Pemberton's and that, to do so, they must crush the Union force at Clinton between them. That very night, the night of the 13th, Johnston sent the following order to Pemberton by courier:

Jackson
May 13 8:40 P.M.
Lt General Pemberton
   I have lately arrived and learned that Major General Sherman is between us, with four divisions at Clinton. It is important to re-establish communications, that you may be re-enforced. If practicable, come up in his rear at once. To best such a detachment would be of immense value. The troops here could cooperate. All the strength you can quickly assemble should be brought. Time is all important.
   J. E. Johnston[5]

Pemberton did not receive the order until between 9 and 10 A.M. on the 14th while he was en route from Bovina to his army encampment at Edwards Station. Thus, over 12 hours elapsed before Pemberton even received Johnston's order, and when he did receive it, he was not with, but en route to, the troops that Johnston wanted to make the attack on Clinton.

President Davis was later to criticize Johnston for sending this order to Pemberton via courier, rather than riding himself to Pemberton's headquarters after arriving at Jackson. In Davis's eyes, a personal conferring with Pemberton could have avoided any confusion.[6]

However, this would have required the ailing 56 year old Johnston to make a 30-mile overnight horseback ride, immediately after completing a fatiguing, stop-and-start, three-and-a-half-day train ride.

Upon receiving General Johnston's order on the morning of the 14th, Pemberton immediately replied by courier, "I have the honor to acknowledge the receipt of your communications. I move at once with whole available force (about 10,000) from Edwards Depot..."[7]

After sending his response to Johnston, Pemberton continued on his ride to his army's encampment at Edwards Station. By the time he completed the ten or so remaining miles, he decided that maybe he should consult with his generals before deciding to make the attack on Clinton, and so he called a council of war. The hours continued to tick away as Pemberton's generals argued and discussed. The majority finally decided that they should make the attack as Johnston ordered. By this time, however, Pemberton decided that it would be better if he did not make the attack. A better course, in his mind, was to proceed south, rather than east to attack Clinton and join Johnston, and thus interpose between the Union forces at Clinton and Raymond and their supply base on the Mississippi. His new object was Dillon to the south rather than Clinton.

It was late on the 14th before Pemberton drafted a message to be sent by courier to Johnston informing Johnston of Pemberton's decision. The message was as follows:

Hdqtrs Dept. Miss. and E. La., Edwards Depot, May 14 1863
General Joseph E Johnston
I shall move as early to-morrow as practicable with a column of 17,000 men to Dillon's, situated on the main road leading from Raymond to Port Gibson, 7½ miles below Raymond and 9½ miles from Edwards Depot. The object is to cut the enemy's communications and to force him to attack me, as I do not consider my force sufficient to justify an attack on enemy in position or to attempt to cut my way to Jackson...
J. C. Pemberton[8]

Pemberton did not actually leave Edwards Station for Dillon until late on the 15th, almost 48 hours after Johnston's order to attack Clinton "at once" with the final caveat, "Time is all important." And, of course, when he did leave, it was south toward Dillon rather than east toward Clinton. When he bivouacked for the night of the 16th, he was only four miles from Edwards Station, his starting point.[9]

General Grant was not idle while Pemberton dallied at Edwards Station trying to decide whether or not to obey Johnston's order. The Union troops at Clinton and Raymond moved on Jackson, and by the afternoon of the 14th, had forced Johnston out and captured the city. Johnston successfully extricated his troops, but was forced to retreat to the north toward Canton (see map 21).

Johnston considered that the situation had changed so radically since he sent his order of 8:40 P.M. on the 13th, that he must issue a new order. The new order was handed to a courier late on the afternoon of the 14th and was as follows:

Camp 7 miles from Jackson
May 14, 1863
Lt General Pemberton
GENERAL: The body of troops mentioned in my note of last night compelled Brigadier-General Gregg and his command to evacuate Jackson about noon today. The necessity of taking the Canton Road at right angles to that upon which the enemy approached prevented an obstinate defense. A body of troops, reported this morning to have reached Raymond last

night, advanced at the same time from that direction. Prisoners say that it was McPherson's corps (four divisions) which marched from Clinton. I have no certain information of the other; both skirmished very cautiously. Telegrams were dispatched when the enemy was near, directing General Gist to assemble the approaching troops at a point 40 or 50 miles from Jackson, and General Maxey to return to his wagons, and provide for the security of his brigade, for instance by joining General Gist. That body of troops will be able, I hope, to prevent the enemy at Jackson from drawing provisions from the east, and this one may be able to keep him from the country toward Panola. Can he supply himself from the Mississippi? Can you not cut him off from it, and above all should he be compelled to fall back for want of supplies, beat him? As soon as the re-enforcements are all up, they must be united to the rest of the army. I am anxious to see a force assembled that may be able to inflict a heavy blow upon the enemy. Would it not be better to place the forces to support Vicksburg between General Loring and that place, and merely observe the ferries so that you might unite, if opportunity to fight presented itself? General Gregg will move to Canton tomorrow. If prisoners tell the truth, the forces in Jackson must be half of Grant's army. It would decide the campaign to beat it, which can be done only by concentrating, especially when the remainder of the eastern troops arrive — they are to be at least 12,000 or 13,000.

J. E. Johnston[10]

This long, chatty, complex message was really not an order at all. If we may digest its essence, it said:

1. The Union troops at Clinton and Raymond attacked and captured Jackson on the 14th.
2. The Confederate troops escaped and are retreating northward on the road to Canton.
3. The reinforcements under General Gist and General Maxey, who were expected to join Johnston at Jackson on the 14th, had not arrived before the loss of Jackson.
4. Gist, who was coming by rail from the East, was directed to debark 40 to 50 miles east of Jackson. Maxey, who was coming up from Port Hudson, was directed to join Gist.
5. It was suggested that Pemberton interpose his force between Grant's troops in Jackson and Grant's supply base on the Mississippi at Grand Gulf. The words were, "Can you not cut him off from it?"

Strangely enough, the action that Johnston now suggested that Pemberton undertake was precisely what Pemberton was attempting to do by his march to Dillon. In any event, it was all for naught, inasmuch as Pemberton did not receive Johnston's message until 6 P.M. on the 16th, over two days after it was handed to a courier. By that time, the issue had been decided.

On the morning of the 15th, Johnston finally received Pemberton's dispatch that advised him that Pemberton was moving on Dillon and not on Clinton. This was more than 12 hours after Pemberton had turned the dispatch over to a courier. Johnston replied as follows:

Canton Road
Ten Miles from Jackson
May 15 8:30 A.M.
Lt General Pemberton, Commanding etc.

Your dispatch of yesterday just received. Our being compelled to leave Jackson makes your plan impracticable. The only mode by which we can unite is by your moving directly to Clinton, informing me, that we may move to that point with about 6000. I have no means of estimating the enemy's force at Jackson. The principal officers here differ very widely. I fear

he will fortify it if time is left him. Let me hear from you immediately. General Maxey was ordered back to Brookhaven. You probably have time to make him join you. Do so before he has time to move away.
  Most Respectfully,
  J. E. Johnston[11]

Pemberton received Johnston's message 22 hours later and replied as follows:

Four Miles South of Edwards Depot,
May 16, 1863
General Joseph E. Johnston,
  Your letter, written on the road to Canton, was received this morning at 6:30. It found this army on the middle road to Raymond. The order to countermarch has been issued. Owing to the destruction of a bridge on Baker's Creek, which runs for some distance parallel with the railroad and south of it, our march will be on the road leading from Edwards Depot in the direction of Brownsville. This road runs nearly parallel with the railroad. In going to Clinton we shall leave Bolton Depot 4 miles to the right. I am thus particular, so that you may be able to make a junction with this army.
  J. C. Pemberton[12]

Two things are noteworthy about Pemberton's reply. First, he indicated his location as four miles south of Edwards Depot. Thus, in the three days since Johnston wrote his original order of the 13th to Pemberton, Pemberton had only moved four miles. Second, we must note the alacrity with which Pemberton obeyed this order. This is in contrast with his dilatory response to Johnston's first order to move to Clinton. In that instance, he first called a council of war and then decided to move on Dillon and not Clinton. The difference in response might be explained thus: although this was Johnston's third order, the second one recommending that Pemberton move on Grant's line of communications had still not been received. Pemberton had only received two of Johnston's orders, both to proceed to Clinton. Thus, Pemberton may have concluded that Johnston was angry with him for not doing so. Pemberton, however, was destined never to get to Clinton and never to unite with Johnston.

When Johnston drafted his third order to Pemberton, he gave it to three couriers to ensure delivery. One of the three happened to be a Union spy.[13] The courier delivered the message to Union corps commander McPherson who, in turn, delivered it to Grant. Grant read Johnston's order to Pemberton before Pemberton read it. Grant, with his customary decisiveness, gathered his forces to meet Pemberton before Pemberton could unite with Johnston.

Grant hit Pemberton at Champion's Hill, nine miles west of Clinton, on the morning of May 16 (see map 21). The Confederates put up a stout defense, but numbers ultimately prevailed. Grant brought 32,000 men into the battle versus Pemberton's 22,000. By darkness it was all over and Grant was victorious in the key battle of his Vicksburg campaign.

Pemberton retreated westward with the bulk of his army toward the Big Black River, which was just before the main defenses of Vicksburg. Here, he prepared to make another stand. General Loring's division of Pemberton's army was cut off from the main body in the battle, but ultimately made its way to join Johnston.

Johnston knew nothing of the battle of Champion's Hill, or of Pemberton's retreat and defeat as it transpired. He was awaiting Pemberton's response to his order to move to Clinton, at his encampment on the Canton Road. Pemberton's response finally came as darkness fell on the 16th. It indicated that he was moving on Clinton as ordered and provided information for their rendezvous.

Johnston set out for Clinton and the rendezvous at daylight on the 17th. He had marched 15 miles when he received a message from Pemberton. It proclaimed disaster. It advised Johnston of Pemberton's defeat at Champion's Hill on the 16th, of his retreat to the Big Black, of his further defeat at the Big Black, and his retirement into the environs of Vicksburg. It further advised that, as a result of the retirement from the Big Black, Snyders Mill (Haynes Bluff), the northern anchor of Vicksburg's defenses, was in danger.

Johnston immediately replied to Pemberton with the following message:

> If Haynes Bluff is untenable, Vicksburg is of not value and cannot be held. If therefore you are invested in Vicksburg, you must ultimately surrender. Under such circumstances, instead of losing both troops and place, you must, if possible, save the troops. If it is not too late, evacuate Vicksburg and its dependencies, and march to the northeast.[14]

If Pemberton retreated to the northeast as Johnston desired, Pemberton and Johnston would still be able to unite. Pemberton did not receive Johnston's order of the 17th until about noon on the 18th. Once again, Pemberton, instead of complying with Johnston's desire, decided to call a council of war and determine the opinions of his generals. Upon hearing them out, he dispatched the following message to Johnston:

> Hdqtrs, Department of Mississippi and Eastern Louisiana
> Vicksburg, May 18, 1863
> General Joseph E. Johnston:
>   General: I have the honor to acknowledge the receipt of your communication, in reply to mine by the hands of Captain Henderson. In a subsequent letter of the same date as this latter, I informed you that the men had failed to hold the trenches at Big Black Bridge, and that as a consequence, Snyders Mill was directed to be abandoned. On the receipt of your communication, I immediately assembled a council of war of the general officers of this command, and having laid your instructions before them, asked the free expression of their opinions as to the practicability of carrying them out. The opinion was unanimously expressed that it was impossible to withdraw the army from this position with such morale and material as to be of further service to the Confederacy. While the council of war was assembled, the guns of the enemy opened on the works, and it was at the same time reported that they were crossing the Yazoo River at Brandon's Ferry, above Snyders Mill. I have decided to hold Vicksburg as long as possible, with the firm hope that the Government may yet be able to assist me in keeping this obstruction to the enemy's free navigation of the Mississippi River. I still conceive it to be the most important point in the Confederacy.
>   Very respectfully, your obedient servant.
>   J. C. Pemberton
>   Lt. General, Commanding[15]

This sealed the fate of both Vicksburg and Pemberton's army. Grant quickly threw up entrenchments about the city and sealed Pemberton in. Now it was siege warfare and just a matter of time before the garrison was driven to starvation. Pemberton finally surrendered Vicksburg on July 4, 1863. He had lost both the fortress and his army.

Grant's victory at Vicksburg was perhaps the greatest victory of the war and sealed the fate of the Confederacy. Gettysburg is often cited as the turning point of the war. The South lost nothing at Gettysburg. It merely failed to win the war. The battle was fought entirely on Union terrain. The southern army remained intact and was fully capable of fighting another day.

The damage done to the southern cause at Vicksburg was irreparable. The South was not only rent in two, but it lost an army. The ramifications of the southern defeat went beyond this. Grant now rose to the top in the eyes of the Union political leadership. Had he lost, he would have remained just one of many generals. He was now singled out to head all the Union armies—and because of his leadership, the Union achieved final victory in two years.

An Analysis

Johnston and Pemberton obviously had vastly different conceptions of the problem facing them. Pemberton considered that the holding of Vicksburg was of paramount importance. In Pemberton's post battle official report he stated:

> The evacuation of Vicksburg: It meant the loss of the valuable stores and munitions of war collected for its defense; the fall of Port Hudson; the surrender of the Mississippi River, and the severance of the Confederacy ... I still conceive it to be the most important point in the Confederacy.[16]

Johnston, on the other hand, believed that the preservation of their army and the destruction of Grant's was the main task—even if that necessitated the temporary relinquishment of Vicksburg. If they won, they would get Vicksburg back. If they lost, they would lose both the army and Vicksburg.

Johnston's view was undoubtedly the militarily correct one. In any event, he was a full general and Pemberton a lieutenant general. Johnston was Pemberton's boss. His views should have prevailed. Why was he unsuccessful in imposing his views and will on Pemberton?

From the time Johnston arrived at Jackson on May 13 until Pemberton retired into the walls of Vicksburg on May 17, Johnston sent Pemberton four orders as follows:

1. Sent May 13—Attack the Union forces at Clinton
2. Sent May 14—Cut Grant's supply line to the Mississippi
3. Sent May 15—Rendezvous at Clinton
4. Sent May 17—Retreat to the northeast

Pemberton did not obey (1) or (4) and could not have obeyed (2) because he did not receive it while obedience was still possible. He attempted to obey (3), possibly because, not having received (2), he thought that this was a repeat of (1) from a Johnston who was angry because of Pemberton's failure to execute (1).

As we have previously noted, Johnston telegraphed two orders to Pemberton after learning of Grant's crossing the Mississippi while Johnston was still at Tullahoma. Both orders were to the effect that Pemberton should immediately consolidate all his forces and confront Grant as near the crossing as possible. Pemberton obeyed neither.

Thus, when Johnston arrived at Jackson on May 13, he should have known that he was dealing with a recalcitrant subordinate. With this background, let us look again at the exact wording of the first four orders Johnston couriered to Pemberton after arriving at Jackson:

1. Sent May 13—"If practicable, come up in his rear at once"
2. Sent May 14—"Can you not cut him off from it..." (the Mississippi)

3. Sent May 15 — "The only move by which we can reunite is by your moving directly to Clinton."
4. Sent May 17 — "You must, if possible, save the troops. If it is not too late, evacuate Vicksburg and its dependencies, and march to the north east."

All of the above are conditional orders providing discretion to the recipient. It is highly unlikely that any commander could be convicted by courts-martial for a failure to obey.

Johnston's intent and desire was clearly for Pemberton to attack the Union forces at Clinton as early as possible on the day of May 14. Let us suppose Johnston's order dropped the words "if practicable," and read, "You will attack the Union forces at Clinton with all the forces available to you immediately upon receipt of this message." This would have provided Pemberton with no wiggle room. He would have had to attack.

What then would have transpired? The answer is that we simply do not know. Presumably, Pemberton would have attacked. However, of necessity, the attack would have been delivered far later than Johnston visualized. Pemberton was not handed Johnston's order until between 9 and 10 A.M. on the 14th. When he did receive it, he was not with his troops who were to make the attack, but was ten miles away at Bovina. He could not have reached his troops at Edwards Station and ordered the attack before 10 or 11 A.M. To add a complicating factor, there were torrential rains in the Jackson area on the morning of the 14th. Although there were too many variables to predict the outcome of a Pemberton attack on the morning of the 14th, one thing we can say with certainty. It could not have worked out worse for the Confederacy than it actually did by him not making the attack. The odds for him were better on May 14 than they were for him when he finally did fight the battle, not at Clinton, but at Champion's Hill on May 16.

Now let us turn to Johnston's order to Pemberton on May 17, the one ordering him to evacuate Vicksburg and march to the northeast. Suppose Johnston eliminated the words "if possible" and "if it is not too late," and the order read, "You will evacuate Vicksburg immediately upon receipt of this order and march to the north east." Had this been the order, Pemberton would have had no choice but to obey, and his army would probably have lived to fight another day — and would probably have united with that of Johnston.

Before concluding our analysis, let us look at the timeliness of the communications between Pemberton and Johnston. By the time Johnston arrived in Jackson on May 13, Union troops were on the rail line between Jackson and Vicksburg, and had cut the telegraph wire. Thus, all communications between Johnston and Pemberton had to be by courier. Between the time Johnston arrived on the 13th and Pemberton holed up in Vicksburg on the 17th, they were never more than 30 miles apart, and sometimes less than 20. It took a minimum of 12 hours and sometimes more than 48 hours to deliver a message. Thus, at the absolute minimum, when Johnston sent an order to Pemberton, he did not receive a response for 24 hours. Considering that they were acting in a fast-moving situation, it is not surprising that their two forces failed to unite.

What Happened to Them

After the loss of Vicksburg, it was readily apparent that Pemberton, a Yankee in Rebel clothing, could never again be entrusted with a command commensurate with his rank. It was politically impossible because he was widely disliked and distrusted throughout the South. Consequently, he remained without assignment. Pemberton, however, anxious to prove his loyalty to the South, petitioned Davis to reinstate him in any capacity whatsoever. Davis, who never lost trust in Pemberton, reinstated him in the rank of lieutenant colonel. Pemberton accepted and faithfully and competently served in this rank until the end of the war.

After the war, Pemberton engaged in farming for a time in Virginia, but ultimately returned to his roots in Pennsylvania. He died in Philadelphia in 1881 at the age of 67. The locals objected to his being buried in Laurel Cemetery, considering him a traitor. However, he was ultimately buried in a remote part of the cemetery, and there he remains.

Joseph Johnston was later to assume the command of the other army of his domain, that of General Bragg. Here, he again infuriated President Davis with his proclivity to retreat. When Atlanta became threatened, Davis foolishly replaced Johnston with John Bell Hood. Hood quickly led the Confederate forces to disaster.

Davis, later, at Lee's instigation, replaced Hood with Johnston. However, it was too late and the end was in sight. Johnston performed admirably, but it was all over.

Johnston had a successful postwar career serving as insurance executive, congressman, and commissioner of railroads. He died in 1887 at the age of 84.

Johnston remains the most controversial of all the Confederate generals. In the view of some, he was the only one who had the keys to victory. In the eyes of others, he is the one most responsible for defeat.

10

Gettysburg

Our next case is probably the most important of all, and is also different from the others. The case is the battle of Gettysburg. The conventional wisdom is that the Confederates just might have won had it not been for two faulty battlefield orders, one to General Ewell and one to General Stuart. It is usually accepted that in each of these instances, the order contained too much discretion and that, in each instance, the recipient acted contrary to the will of the originator. Had the orders been otherwise, and had the recipient taken the intended action, there is good reason to believe that the Confederates would have won. The ramifications for our history are enormous.

Had the Confederates won at Gettysburg, the likelihood is great that Lincoln would have lost the election of 1864 and George B. McClellan would have become president. Had McClellan become president, the chances are great that the war would have ended in a negotiated settlement. Had this occurred, we would be two nations today instead of one.

Let us begin by examining the events that led up to the battle of Gettysburg. Then, we will dissect the crucial first day of the battle. Next, we will look at the orders in question and the circumstances surrounding their issuance and implementation. Finally, we will decide as to whether or not a slight change in the wording of these orders could have resulted in a totally different outcome.

The Events Leading Up to Gettysburg

Between May 1 and 4, 1863, Lee and his Army of Northern Virginia inflicted a stunning defeat on the Union Army of the Potomac under Hooker, at the battle of Chancellorsville. Most stunning of all, Lee had done it with an army half the size of that of Hooker. Lee's army of the time normally consisted of two corps, headed by Longstreet and Jackson, plus cavalry. At the time of the battle, Longstreet's corps was away on detached duty in southern Virginia. Lee had, in effect, triumphed over Hooker with one arm tied behind his back. If he could beat the Union army with one arm, what could he accomplish with two?

Longstreet rejoined Lee and Lee now considered that it was time to take the offensive. He would invade the North for the second time. Lee's army was not only far larger, better organized, and better equipped than during his first invasion, but now, morale was sky high.

Both the men in the ranks and the high command were infused with a feeling of invincibility, that one Rebel could beat two Yankees. And why not? The rebels had triumphed over larger Union armies at First Bull Run, Second Bull Run, Fredericksburg, and now Chancellorsville.

There was, however, one little problem. "Stonewall" Jackson, the almost legendary figure, was dead. He was wounded at Chancellorsville and died soon after. There was no one of his caliber to replace him. Consequently, Lee reorganized the army from two corps to three. Longstreet continued to head the First Corps and remained second in command to Lee. Lieutenant General Ewell was given command of the new Second Corps, and Lieutenant General A. P. Hill was given command of the new Third Corps. General Stuart remained in command of the cavalry.

When comparing Lee's plans for his first and second invasions of the North, we note

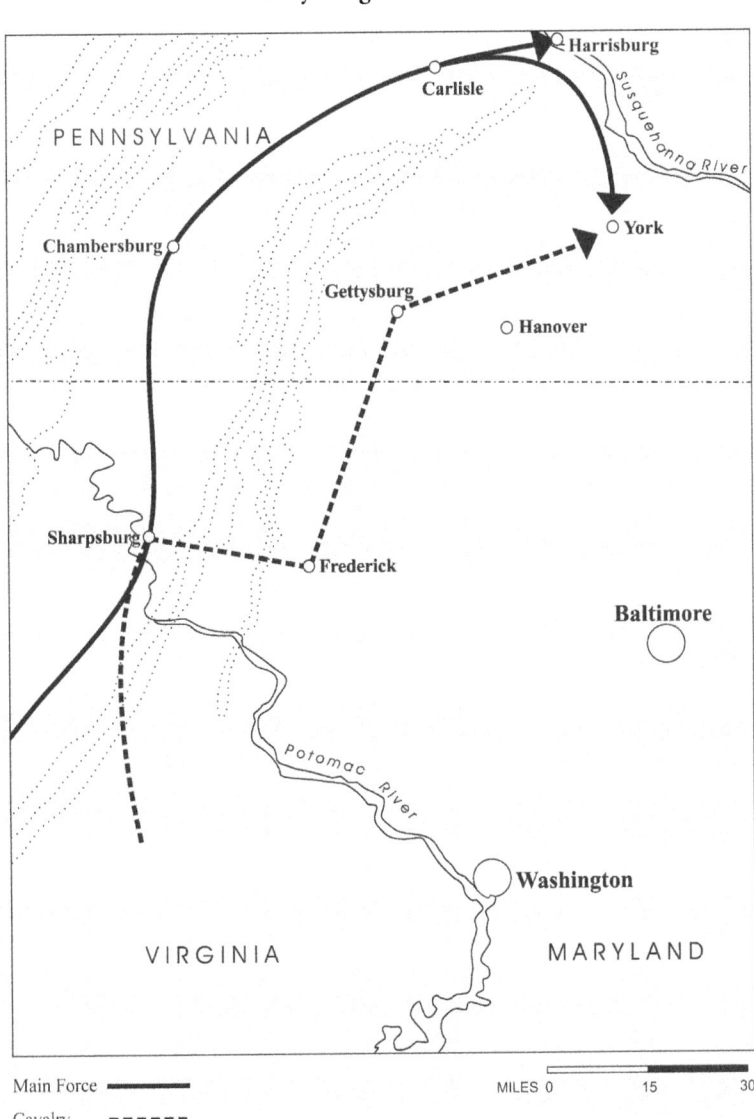

Map 22
Gettysburg — Lee's Plan

both similarities and differences. In both instances, he planned to move up the Shenandoah and Cumberland Valleys through Maryland and into Pennsylvania. This would have multiple objectives. First, he would cut three of the four lifelines connecting the east of the Union to the west. These were: the Baltimore and Ohio Railroad, which was just south of the Potomac; the Chesapeake and Ohio Canal, which was just north of the Potomac; and the Pennsylvania Railroad, which crossed the Susquehanna River at the mouth of the Cumberland Valley. Next, he would seize the Pennsylvania capital of Harrisburg. Lee considered that seizing a Northern capital, plus defeating the Union army in battle on its own soil, would secure foreign recognition of the Confederate states as a sovereign country.

There was one major difference in the plans for the first and second invasions. In the first invasion, the Confederates crossed the Potomac at Leesburg, well to the east of the Shenandoah Valley. It was in their attempt to move the invasion into the valley while north of the Potomac that the invasion went awry and ended abortively in the battle of Antietam. In the second invasion, the invading army would enter the Shenandoah Valley while still in Virginia.

One great advantage of Lee's plan was that, as the Confederate army moved up the valleys, the Union commanders could never be sure of his objective. At any time the Confederate army might debouch from the valley and head for Washington, or Baltimore, or Philadelphia. The Union army would thus have to stay to the east to keep between him and these cities.

Specifically, Lee's plan for the second invasion was as follows (see map 22): Ewell, with the Second Corps, would enter the valley first and proceed north as the point of the invading army. His corps would clear out any of the small Union garrisons in the valley. Ewell's corps would be followed by Hill's, and Longstreet's corps would bring up the rear. Stuart's cavalry would hold the passes into the valley until the invading army was safely across the Potomac. The bulk of the cavalry would then proceed north past the moving army and join Ewell at the spear point. The point where the cavalry was to join Ewell was York, Pennsylvania.

June 30, 1863

Up to June 30, Lee's plan worked to near perfection. Ewell had cleared the valleys in a series of minor victories and had now reached the Susquehanna River. His corps occupied both Carlisle and York, Pennsylvania. Hill's corps was centered at Chambersburg, and Longstreet's corps was a day behind.

There was, however, one very large problem. Lee had heard nothing from Stuart since Stuart's departure from the rear of the army on June 25 for York. Lee had come to depend upon Stuart and his cavalry for intelligence ever since he had assumed the command of the Army of Northern Virginia. Without Stuart, he was forging ahead blind. As of June 30, the enemy had better information on his movements than he had on theirs.

The positions of Confederate and Union forces as of June 30 are indicated on map 23.

Before we proceed to describe the critical and confusing events of July 1, let us look at the matter of time in the Civil War.

10. Gettysburg

Map 23
Gettysburg — June 30th

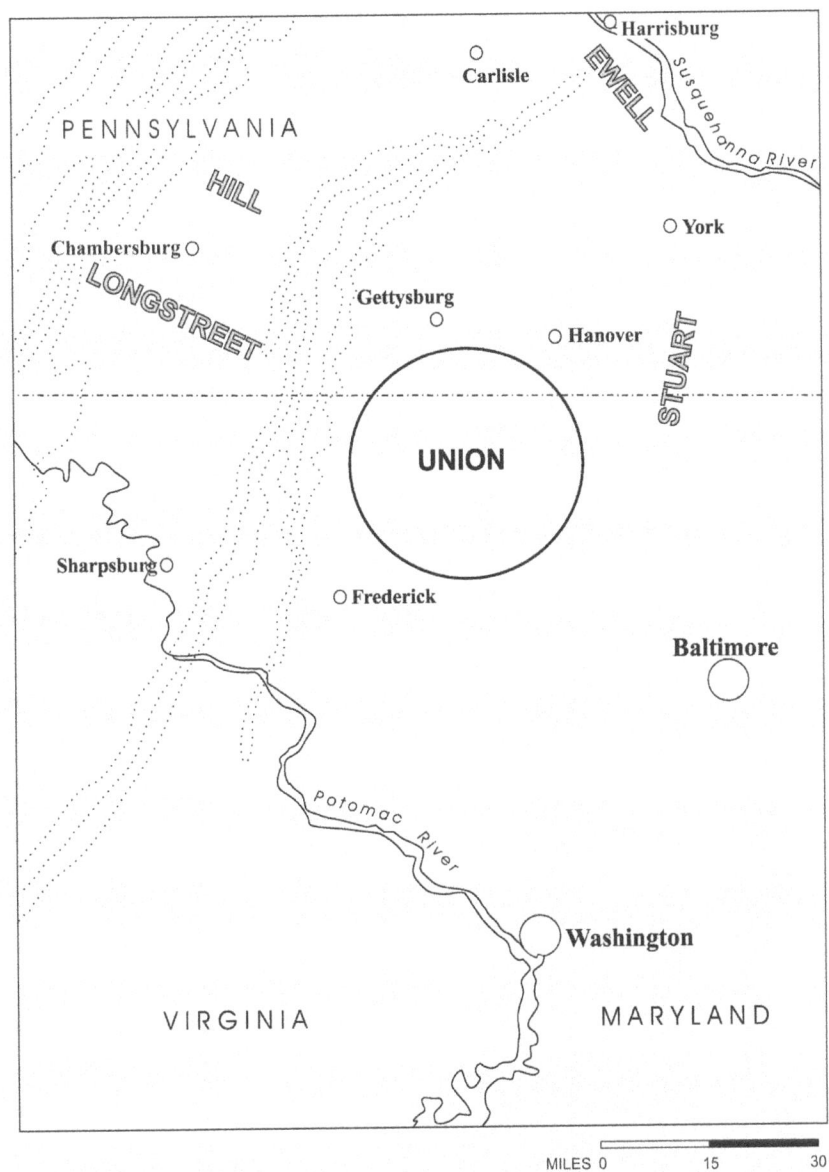

Time in the Civil War

In order to determine the sequence of events, one must know the time that each happened. This poses an enormous problem for Civil War historians. There was no standard time or time zones at the time of the Civil War. Each jurisdiction kept its own time. When the sun was directly overhead, it was considered noon, and the rest was based upon that moment. Inasmuch as the position of the sun relative to the earth is constantly changing, each jurisdiction had a different time and, in the case of a moving army, the watches of officers could be set to different times. Because of this, Civil War writers of post battle reports often used references to the sun rather than hours and minutes when placing events. For example, they might refer to a time as "sunset" or "one hour after sunset" or "noon" or "evening" or "sunrise," etc.

To cite but one example, what time was Culp's Hill occupied by the Union? During the "evening" of July 1, General Hancock ordered General Wadsworth to occupy Culp's Hill. During that "evening," General Wadsworth passed the order to Colonel Grover, and Colonel Grover so ordered the troops. We do not know exactly what is meant by "evening" and we have no idea how much time elapsed between the time General Wadsworth received the order and then passed it on to Colonel Grover, nor do we know how much time elapsed before Grover passed on the order. We do not know exactly where the troops who were ordered to occupy Culp's Hill were, nor when the first one got underway. In all, over 1,100 troops occupied the Hill, and this undoubtedly took time, possibly as much as an hour. Was the hill considered "occupied" when the first soldier arrived, or not until the last arrived, or sometime in between? Not surprisingly then, there is a great disparity in the estimates of various historians as to the time the hill was occupied.

Reports often indicated that corps so-and-so arrived at a certain time. What does this mean? Generals often rode on ahead and reported in before the head of the columns arrived. Then too, it took more than an hour and often more than two hours before the entire corps arrived. Did it arrive at the time the general reported, the time the first man arrived, or the time the last man arrived?

Next, we come to the question of "where" did it arrive. Usually, we have a general term such as "York" or "Gettysburg." Most of the activity at Gettysburg was centered at Cemetery Hill, on the southern outskirts of Gettysburg. Thus, a unit could arrive at the northernmost limits of Gettysburg and still be an hour's march from the scene of the action.

In conclusion, the opportunity for placing events out of sequence is large and, often, the times given for events are rough approximations at best.

The Leading Players

The leading characters of this episode are Confederate Major General J. E. B. Stuart, head of Lee's cavalry, and Lieutenant General Richard Ewell, commanding general of Lee's Second Corps.

Stuart graduated from West Point in 1854, long after the Mexican War was

Major General J.E.B. Stuart CSA: One try too many for "glory."

over. Had it not been for the Civil War, in all likelihood we would know nothing of him today.

Stuart resigned his commission in the U.S. Army and joined the Confederate Army in May 1861. His rise in rank was meteoric. Within just 14 months, at the age of 29, he was a major general in command of all the cavalry in Lee's army.

Despite Stuart's youth, he appeared supremely competent and went from success to success. He gained the confidence of Lee, who depended on him ever more for the provision of accurate and timely intelligence on the enemy.

Perhaps because of his youth, Stuart was flamboyant and appeared to bask in publicity and public approbation. He wore a red-lined grey cape, yellow sash, and hat cocked to the side adorned with an ostrich plume. Each of his feats was trumpeted by the press. One such feat occurred during the Peninsular Campaign when, on an intelligence collection mission, he rode completely around the Union army.

During the war, unit commanders were required to submit written post battle reports outlining their unit's actions. These were for study and historical purposes. Many if not most unit commanders looked upon this as just one more bureaucratic burden and devoted as little time to them as possible. Not Stuart! His reports outlined every little accomplishment that could reflect favorably upon him.

After the smoke cleared at Gettysburg and the South began looking for a scapegoat, many looked at Stuart. Were his actions, which proved so detrimental to the South, motivated by a desire for more fame and publicity rather than the betterment of the cause? We will never know. After the battle, Stuart reverted to his usual competence, and Gettysburg was but an unfavorable blip on an otherwise illustrious performance.

Stuart was not to survive the war. He was mortally wounded at the battle of Yellow Tavern and died on May 12, 1864, at the age of 31.

Our other major player, General Ewell, was about as far in background and personality from Stuart as one could get.

Ewell was 46 years old at the time of Gettysburg, and had graduated from West Point way back in 1840, long before such Civil War luminaries as Grant, Sherman, McClellan, and Sheridan. At the time he resigned his commission and entered the Confederate service in 1861, he had over 20 years of distinguished service. He had seen much combat in both the Mexican and Indian Wars, and consistently showed no reluctance to get in harm's way.

Ewell entered the Confederate service as a colonel and quickly rose to general. He served under Jackson as Jackson's principal assistant in Jackson's fabled Valley Campaign, and always succeeded in maintaining harmonious and constructive relations with Jackson, which was no small feat. Upon Jackson's premature death, Ewell was the logical candidate to fill Jackson's giant shoes.

In the ten months preceding Gettysburg, Ewell was to undergo two life-altering experiences that, some say, affected his personality. First, at the second battle of Bull Run on August 30, 1862, Ewell received a serious wound that necessitated amputating his leg at the knee. At the time, such amputations were horrifying experiences that the subject would naturally try to avoid repeating at all costs. Second, the middle-aged Ewell married for the first time. Be that as it may, Gettysburg, for Ewell, was but one demerit in an otherwise illustrious career. Ewell was in the war from beginning to end and survived to become a gentleman farmer.

Richard Taylor, Confederate lieutenant general and a son of President Taylor, perhaps best sums up Ewell in his memoirs, "Dear Dick Ewell! Virginia never bred a truer gentleman, a braver soldier, nor an odder more loveable fellow."[1]

July 1, 1863

July 1, 1863, was a pivotal day in U.S. history. Actions this day were to determine whether we remained one country or became two.

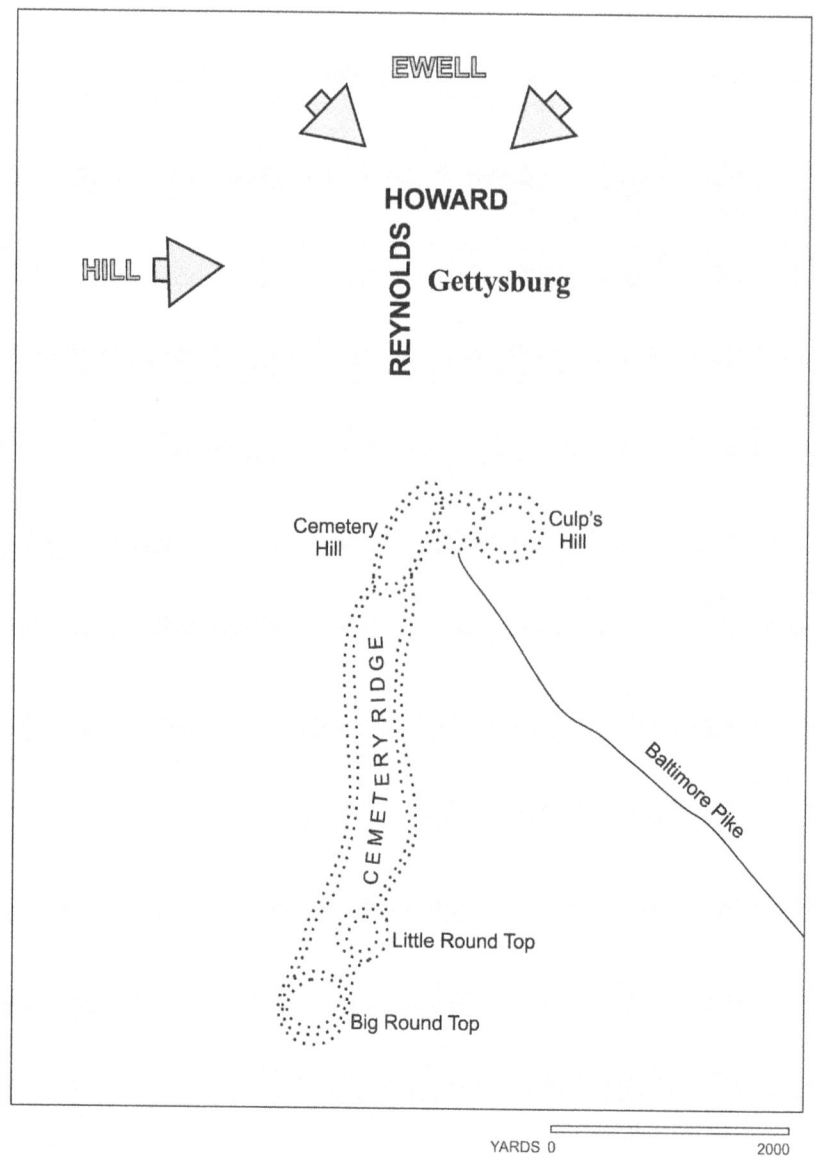

Map 24
Gettysburg — July 1st

As Hill's corps passed the city of Gettysburg, he heard that there was a large supply of shoes there. Shoes were an important asset for a marching army, and many of Hill's men were barefoot. Hill sent one of his three divisions, that of Heth, into town to confiscate the shoe supply. Just outside of town, Heth's troops ran into a Union cavalry brigade under General Buford, and the first shots of what was to become the battle of Gettysburg were fired.

At this juncture, Lee sent an order to his corps commanders. They were to join at Gettysburg, but were not to bring on a general engagement yet. Lee wanted the army united before a general engagement commenced. Lee's order was to have significant ramifications and, in fact, was a bad call on Lee's part. In a meeting situation, the side that can congregate its forces on the scene first generally wins. Unknown to Lee, as of the morning of the first his army was in the best position to congregate his forces first. This advantage, however, was transitory. By the end of the day, Union troops would be arriving on the scene faster than his. Unknown to Lee, his best chance to win would be on July 1. Now was the best moment for him to initiate a general engagement.

General Heth called to General Hill for reinforcements, and General Buford called to General Reynolds, head of the Union First Corps, and to General Howard, head of the Union Eleventh Corps, for help. Whether Lee wanted it or not, de facto, the battle of Gettysburg was on.

As we can see from map 24, the Confederate forces from General Hill's corps were coming in from the west and those of General Ewell's corps were coming down from the north. The Union defending forces thus had to form two defense lines interlocking at a right angle. Howard took the line to defend against Ewell and Reynolds took the line to defend against Hill. By 2 P.M., the numbers of arriving Confederate troops began to overwhelm the Union defenders of both lines, and by 4 P.M., both Union lines were flanked and had reached the breaking point. Reynolds was dead and Howard was now in charge. Howard gave the order for both lines to retreat and to rally at Cemetery Hill, a prominence just south of Gettysburg.

As Howard's troops poured into the narrow streets of Gettysburg from the north and Reynolds's (now Doubleday's) poured into the streets from the west, units became intermingled, cohesion was lost and, with the Confederates firing into the masses, an orderly retirement turned into a rout. We will turn to an eyewitness account:

> The Union troops driven into the town from different directions were wedged and jammed in the streets, and soon became a disorganized mass. Artillery and ambulances struggling to get through the tangled crowd added to the confusion. Had the fugitives been allowed no pause, and had the Confederates followed close on their heels, the very momentum of the flight, to say nothing of the contagion of the panic, would have swept away every support, and the pursuers could easily have rushed the cemetery and the surrounding heights.[2]

If there ever was a time for cavalry pursuit, this was it. But alas, Stuart and his three brigades were not there and Lee had no idea where they were. Had they been there, Lee might have cemented his victory on July 1.

Cemetery Hill, which was just south of Gettysburg, was a natural strong point. When General Howard set up his defense line to the north of Gettysburg, he had purposely left a brigade on Cemetery Hill as a rallying point, just in case he was forced to retreat. Now it paid off in spades. The oncoming Union hordes quickly reformed and reorganized, and the position grew in strength by leaps and bounds. Within hours, it had become so strong that, as subsequent events were to prove, it could not be taken at all—that is, by direct assault.

It was at about this time that the commanders of both armies noticed the importance of Culp's Hill. Culp's Hill was just to the northeast of Cemetery Hill and was slightly higher. In fact, it dominated the position on Cemetery Hill. If occupied by the enemy, the Union position on Cemetery Hill could not be held. At the moment, Culp's Hill was unoccupied. Its price tag was to rapidly rise from "free" to "priceless."

Culp's Hill

As of June 30, 1863, none of the Union or Confederate commanders expected or planned to fight a battle at Gettysburg. Gettysburg was just a name on a map, and none of them knew anything about its topography. In retrospect, it is not surprising that the meeting occurred at Gettysburg. All roads converged on Gettysburg.

As the forces of both armies converged, generals from both sides looked at the terrain with a military eye. If one were on the defensive, one would naturally want to occupy the high ground between the enemy and one's supply base. One would also want "interior lines," so that he could transfer troops from flank to flank more quickly.

The high ground that stood between the Confederates and the Union supply base at Baltimore was a ridge starting at Cemetery Hill and extending 2.7 miles to the southwest and terminating at two hills called Little Round Top and Big Round Top. Once the Union was firmly established on Cemetery Hill, it gradually extended its line to the Round Tops.

However, just to the northeast of Cemetery Hill, there was another hill, a wooded hill that was slightly higher than Cemetery Hill. This was Culp's Hill. It was within both rifle range and cannon range of Cemetery Hill and, if occupied by the enemy, the Union position on Cemetery Hill would be untenable. If the position on Cemetery Hill was untenable, the whole Union line to the Round Tops would be untenable.

Union occupation of Culp's Hill would have other advantages besides protecting their line on the ridge. It would change their defense position from a straight line to a fish hook shaped line. This would provide them with interior lines. Their distance from flank to flank would be less than two miles, while the distance from flank to flank for the opposing Confederates would be over twice as great (see map 24). Lastly, Union occupation of Culp's Hill would give added protection to their major supply line, the Baltimore Pike. The pike ran directly into the shank of the hook.

By mid-afternoon of July 1, both sides realized that they must have Culp's Hill. Let us see what each side did. We will start with the Union.

Union General Hancock arrived at Gettysburg at 3 P.M. on the first to take command of all Union forces thus far engaged. Sometime later, time unspecified, he ordered General Wadsworth to occupy Culp's Hill. This order to General Wadsworth could not possibly have taken place before 4 P.M. and almost certainly not until well after 5 P.M. At 3 P.M., Wadsworth's division was in the center of the Union line, desperately trying to hold off the Confederates west of Gettysburg. At 4 P.M., his division was among the hordes fleeing through the streets of Gettysburg toward Cemetery Hill. Wadsworth, in his post battle report, says, "On the evening of the 1st, we were ordered to occupy a hill on the right of the cemetery."³ So it was "evening" when Wadsworth got the order. Inasmuch as July 1 is less than two weeks from the longest day of the year, 6 P.M. or later would be a better approximation of events than 5 P.M. Wadsworth, in turn, ordered two of his subordinate units to occupy the hill. These were the 7th Indiana and the remnants of the "Iron Brigade." The 7th Indiana, under Colonel Ira Grover, did not arrive at Gettysburg on the first until after the fight was over and, up to its being ordered to Culp's Hill, had suffered no casualties. This again attests to the lateness of the hour at which the order was issued. The Iron Brigade, on the other hand, had been in the thick of the day's fighting, and had suffered over 60 percent casualties. Its general and most of its senior officers had been casualties, and it must have required time after arriving on Cemetery Hill to organize for the move to Culp's Hill. In all, the combined forces of the 7th Indiana and the remnants of the Iron Brigade totaled little over 1,000 men. Consequently, even when the last man arrived on Culp's Hill, the garrisoning of the hill was modest. My estimate is that the last man of the garrison could not have been on the hill much before 7 P.M. Colonel Grover writes in his post battle report that immediately upon arriving on Culp's Hill, they began constructing a breastwork.⁴

So we have it that the Union occupied Culp's Hill with a small force at around 7 P.M. Now let us see what the Confederates did relative to Culp's Hill.

General Ewell's corps pushed the Union troops that were defending from the north into the streets of Gettysburg in their flight to Cemetery Hill between 4 and 5 P.M. General Ewell himself arrived in the city at about 5 P.M.⁵ At the time, Brigadier General Isaac Trimble was serving on Ewell's staff. Trimble was convalescing from a wound and had not yet returned to duty and received an assignment. Trimble wanted to continue, wanted to seize Culp's Hill immediately. Ewell demurred. He mentioned Lee's order not to bring on a general engagement. He argued that he had been ordered to Gettysburg and no further. In any event, nothing was done and Trimble walked out in anger.⁶

Lee reached Hill's battlefield to the west of Gettysburg at around 4 P.M. He viewed the situation through field glasses and could see disorganized Union troops fleeing through Gettysburg and climbing Cemetery Hill. He wanted to capitalize on the situation and sent his AAG, Walter Taylor, to take a message to General Ewell. Here we will turn to Taylor's own words:

> General Lee witnessed the flight of the Federals through Gettysburg and up the hills beyond. He then directed me to go to General Ewell and to say to him that, from the position which he occupied, he could see the enemy retreating over those hills, without organization and in great confusion, that it was only necessary to press "those people" in order to secure possession of the heights, and that if possible, he wished them to do this. In obedience to these instructions, I proceeded immediately to General Ewell and delivered the order of General Lee; and after receiving from him some message for the commanding general in regard to the prisoners captured, returned to the latter and reported that his order had been delivered.
>
> General Ewell did not express any objection, or indicate any impediment, to the execution of the order conveyed to him, but left the impression upon my mind that it would be executed.⁷

Ewell commanded three divisions. Those of Early and Rodes had been engaged throughout the afternoon. His third division, that of Edward Johnson, had not yet arrived and had seen no action and suffered no casualties. Ewell thought the arrival of Johnson was imminent, and decided that he would give the task of occupying the hills to Johnson. Johnson was delayed and arrived later than anticipated. By the time Johnson deployed his 6,000 man division into battle formation at the base of Culp's Hill, it was already starting to get dark. Johnson sent up a reconnoitering party to see if the hill was occupied. It ran into the small Union garrison that was in the process of digging in, and a shoot-out occurred.

The party returned and erroneously reported that the hill was strongly held. This was, of course, incorrect, but was what the Union garrison wanted the party to report.

In the words of Colonel Ira Grover, the commander of the Union forces on the hill, "During the succeeding night, a force of the enemy attempted to penetrate our lines but was easily driven off, supposing themselves confronted by a heavy force."[8]

By the time the reconnoitering party returned, complete darkness had fallen, and the Confederate commanders, unwilling to undertake what they believed to be a major battle in total darkness in a wooded area, called off any assault that night. By morning, it was too late. The Union forces on the hills and ridge had been reinforced by the Third and Twelfth Corps during the night. Culp's Hill was never to be taken by the Confederates.

In conclusion, Ewell could have had Culp's Hill for little to nothing up to almost 7 P.M. on the first. He could have taken it for a price up to the early morning hours of the second. He would have taken it had Lee's order not been discretionary, containing the two words "if possible." Later, there was some dispute as to exactly what the two words were. Some say "if possible"; others, "if feasible." However, these mean substantially the same thing, and there is no dispute that the order was discretionary.

Now, let us turn to General Stuart and his three cavalry brigades.

Stuart and His Three Brigades

Stuart received his orders in a driving rain storm at Rectors Crossroads, Virginia, on the night of June 23. It called for him to leave two of his five brigades behind to follow the army, and to take the remaining three past the moving army and join the lead corps of Ewell at York, Pennsylvania. Stuart selected the brigades of Robertson and Jones to remain behind, and the three of Fitz Lee, Hampden, and Chambliss to go to York. York was 90 air miles away and, theoretically, Stuart could cover the distance in three or four days.

Stuart set out at 1 A.M. on the 25th. In order to make haste, he took no wheeled vehicles other than ambulances and six pieces of artillery and their caissons.

Instead of proceeding up the valley or alongside the mountains just outside the valley, he decided to proceed through a gap in the Union army before heading north and, in so doing, inflict as much damage as possible on the Union supply lines while en route.

As Stuart started out for York on the early morning of the 25th, the Union army was to his east and south and not yet moving north. He planned to move east, cut through the Union army at a gap discovered by Mosby, and then move north, crossing the Potomac ahead of the Federals, and then on to York. However, before he reached the point where he intended to pass between stationary Union forces, the Union army had taken to the roads and begun moving north, and Stuart found the intended pass-through place solid with marching Union troops. He was forced to proceed ever farther south, away from York, in order to pass around the tail of the Union column. Thus, two days after starting, before he was able to come around to the east of the Union troops and head north, he was further from York than when he started. He finally reached the Potomac at Browsers Ford, 20 miles west of Washington, on the 28th. Over three days had now elapsed; he was still about as far from York as when he started; the Confederate Army was in the valley over 50 miles to the west; and the Union army, proceeding parallel to but behind the Confederate army, was between him and the Confederate army (see map 25).

Stuart finally crossed the Potomac at Browsers Ford on the 28th. Once across the Potomac, he headed north toward York. Upon crossing the Rockville Pike, he encountered

**Map 25
Gettysburg — Stuart's Route**

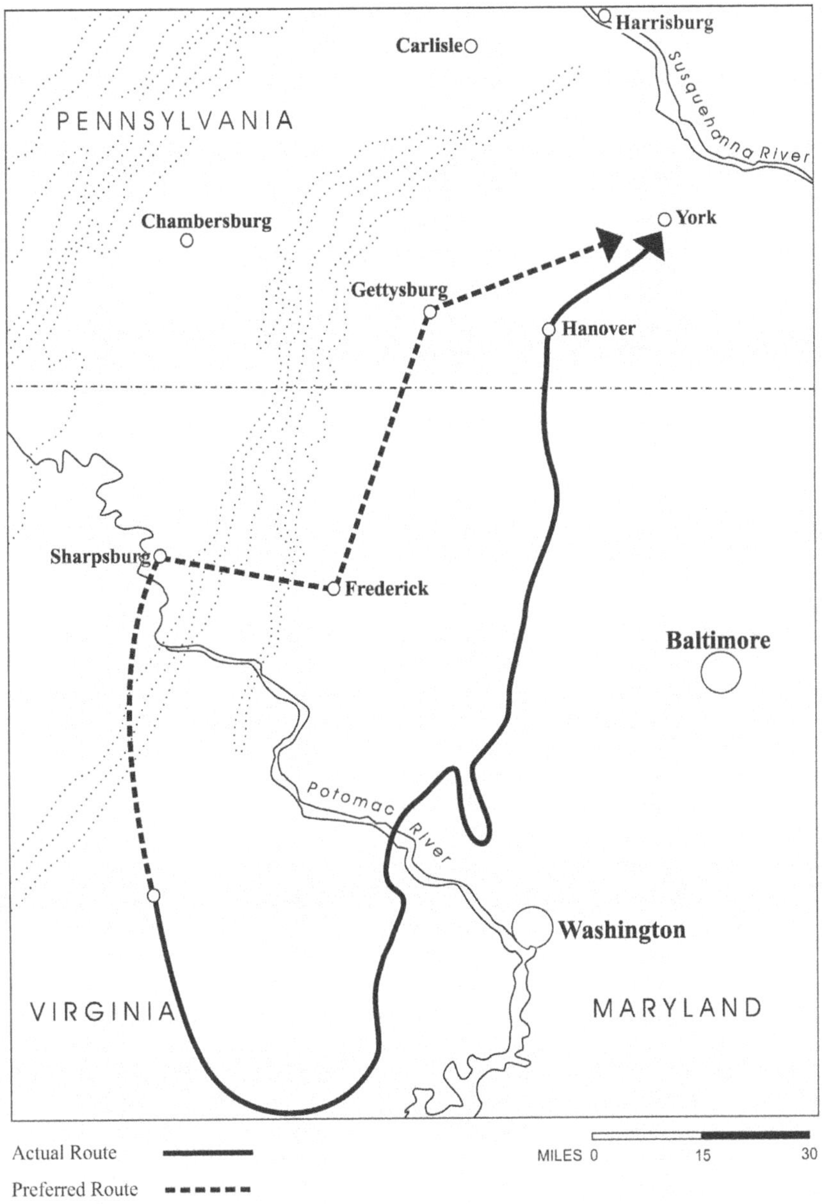

a Union wagon train, lumbering toward the Union army at Frederick. Stuart attacked and captured 125 wagons with their teams. The wagons were mostly loaded with forage. At this point, he made a decision of far-reaching consequences. Even though he had forsaken taking his own wagons for the sake of speed, he now decided to take the 125 wagons with him to deliver to the Confederate army at York.

As Stuart moved north through Union territory, he did inflict some damage on the enemy's supply lines. He destroyed a lock in the Chesapeake and Ohio Canal and several canal boats. He cut the telegraph wires linking the Union army with Washington, and he destroyed a bridge on the B & O Railroad. However, these were mere pin pricks to the Union effort and of little significance. As Stuart proceeded northward, he ran into brushes with the Union cavalry, and what might be called a real battle at Hanover, Pennsylvania.

Stuart finally arrived at York on July 1 but, alas, there were no Confederate forces there. Unknown to Stuart, Lee had called Ewell and all his forces to Gettysburg, and the battle of Gettysburg was already in progress.

Not finding Ewell at York, Stuart proceeded on to Carlisle, Pennsylvania. Here, to his surprise, he found not Ewell but Union militia. Finally, at Carlisle, a courier from Lee found him and told him he was needed at Gettysburg. Stuart finally reported in to Lee at Gettysburg, seven days after he left Rectors Crossroads, Virginia. Stuart presented his captured wagons to Lee. Lee thus received a load of horse feed in place of a victory at Gettysburg.

Here, we will address two vital questions: First, did Stuart violate the orders he received on June 23, and second, did his actions between June 25 and July 2 conform to Lee's desires and intentions? If we find that he did not violate his orders and, at the same time, did not implement Lee's intentions, the orders must have been defective. Let us start by looking at the orders.

> June 23, 1863 5 P.M.
> Major General J. E. B. Stuart,
> Commanding Cavalry
> General:
>
> Your notes of 9 and 10:30 A.M. today have just been received. As regards the purchase of tobacco for your men, supposing that Confederate money will not be taken, I am willing for commissaries or quartermasters to purchase this tobacco, and let the men get it from them, but I can have nothing seized by the men. If General Hooker's army remains inactive you can leave two brigades to watch him, and withdraw the three others, but should he not appear to be moving northward, I think you had better withdraw this side of the mountains tomorrow night, cross at Shepherdstown next day, and move over to Frederickstown. You will, however, be able to judge whether you can pass around their army without hindrance, doing them all the damage you can, and cross the river east of the mountains. In either case, after crossing the river, you move on and feel the right of Ewell's troops, collecting information, provisions, etc. Give instructions to the commander of the brigades left behind to watch the flank and rear of the army, and in the event of the enemy leaving their front, to retire from the mountains west of the Shenandoah, leaving sufficient pickets to guard the passes, and to bring in everything clear along the valley, closing upon the rear of the army. As regards the movements of the two brigades of the enemy moving towards Warrenton, the commander of the brigades to be left in the mountains must do what he can to counteract them, but I think that the sooner you cross into Maryland after tomorrow the better. The movements of Ewell's Corps are as stated in my former letter. Hill's First Division will reach the Potomac today and Longstreet's will follow tomorrow. Be watchful and circumspect in your movements.
> R. E. Lee General[9]

First of all, we can say that the order is not easy to understand, particularly so when received in the middle of the night in a driving rain storm.

We can state categorically that Stuart did not violate his orders. The orders provided him with the discretion of crossing the Potomac at Shepherdstown or passing around (actually through) the Union army and crossing east of the mountains. Stuart selected the latter and only failed because the Union army took to the road and began moving north between the time Stuart received the order and the time he set out to implement it. Had he started one day earlier, he probably would have succeeded in passing through the stationary Union army.

The next question is, did he comply with Lee's intentions? The answer to this is probably "No." Lee's preference, as expressed in the order, was that Stuart pass into the valley, cross the Potomac at Shepherdstown and then, once north of the Potomac, proceed back out of the valley and continue north to York. This was the safe course of action. Shepherdstown, being in the valley, was entirely under Confederate control, and there is absolutely no doubt that Stuart could have crossed the Potomac one day after starting out with absolutely no intervention.

Lee's preferred route, as expressed in the order, was for Stuart to come back out of the valley, after crossing the Potomac, at Frederickstown (presently Frederick) before heading north to York. Here, it was necessary to give Stuart some discretion. It was known that Frederickstown was not occupied by the Union at the time of issuance of the order but, by the time it was delivered, the Union army was moving north toward it.

We conclude that, although Stuart did not violate his orders, he did not comply with Lee's preferences. Let us redraft the order so that he would have been required to comply with Lee's preferences.

June 23, 1863 5 P.M.
Major General J. E. B. Stuart
Commanding Cavalry

1. Upon receipt of this order you will designate two brigades of cavalry to remain behind the army, the senior officers of which will report to General Longstreet.
2. You will take three brigades, designated by you, proceed into the valley and cross the Potomac at Shepherdstown.
3. Once across, you will proceed to York, Pennsylvania, taking such route as circumstances and your judgment dictate.
4. You will report to the senior officer of the Second Corps at York not later than midnight, June 29.
5. En route to York you will: a) screen the army, b) keep in contact with me by a minimum of one report each twenty-four hours, c) provide me with maximum information on the location and movements of the enemy, and d) inflict such damage on the enemy as circumstances warrant.

    I am very respectfully....

Had the above order been provided to Stuart on June 23, there is a good chance that Lee would have won the battle of Gettysburg.

11

Chickamauga

For our next case we will return to the theater of Tennessee, and to its greatest battle, the battle of Chickamauga, Georgia. Shortly after the battle of Perryville, Lincoln removed General Buell from command of the Army of the Cumberland for a lack of aggressiveness, and replaced him with General William Starke Rosecrans. The Confederate opponent remained the same — General Bragg, commander of the Army of Tennessee.

Subsequent events seemed to justify Lincoln's selection of Rosecrans. Largely by masterful maneuvering and deception, he completed the conquering of all of Kentucky and all of western and central Tennessee.

By the summer of 1863, Bragg's main remaining stronghold was Chattanooga and its surrounding territory. Chattanooga was an important railroad junction sitting astride rail lines to the east and west, and to the north and south. As of the summer of 1863, its main supply line was the rail line to the south to Atlanta.

Chattanooga was militarily a tough nut to crack. To its north was the broad Tennessee River. To its south were a series of parallel mountain ridges that ran roughly northeast-southwest.

Rosecrans set out in the summer from his forward supply base at Nashville for what he hoped would be his final campaign to complete the conquest of Tennessee, the capture of Chattanooga.

At this point, the Confederates decided upon a stratagem that had served them well in the past. Foreseeing a final climactic battle for Tennessee, they, always numerically inferior, would at the last minute transfer massive reinforcements by rail to Chattanooga from other theaters, giving them temporary numerical superiority for the actual battle. Specifically, they would transfer two divisions from Mississippi; one from the Knoxville, Tennessee area; and Longstreet's entire corps from Lee's army in Virginia. This, in all, would augment Bragg's force of some 40,000 with an additional 30,000. The last scheduled to arrive was Longstreet's corps, which would not arrive until the third week in September.

The logical thing for Rosecrans to do was to pass Chattanooga on the north side of the Tennessee River, cross over to the east of Chattanooga, and fight the battle for Chattanooga to the east. Rosecrans took extraordinary means to convince Bragg that this was what he was about to do. However, he deceived Bragg once again. While one of his corps, that of Crittenden, passed Chattanooga on the north side of the Tennessee, the other two, those of Thomas and McCook, crossed over to the west of Chattanooga and plunged into the mountain passes to Chattanooga's south.

Bragg sensed that his communications to Atlanta were about to be cut, and evacuated

Chattanooga to the south. Crittenden occupied Chattanooga, and it appeared that Rosecrans had done it again. He had won another bloodless battle and his prestige was never higher.

Here, however, we come to a profound turnabout in fortunes. Rosecrans, with Bragg's assistance, came to believe that Bragg was making a pell-mell retreat to the south and would not make a stand short of Rome, Georgia. He thus concluded that speed of pursuit was essential. To this end he had his three corps take three separate routes through the mountains that resulted in their wide separation (see map 26). Bragg, contrary to Rosecrans's beliefs, was not rushing south in a disorganized fashion. He had retreated only 25 miles southeast of Chattanooga to Lafayette, Georgia, where his army was concentrated. When Rosecrans's three corps debouched from the mountain passes, Bragg's concentrated army was between the northernmost corps of Rosecrans (Crittenden's) and the center corps (Thomas's). The distance between Rosecrans's three corps was: Crittenden to Thomas, 15 miles; Thomas to McCook, 20 plus miles. Rosecrans was in trouble — big trouble.

It was now September 9. Bragg had a window of opportunity. He could destroy either of the northern two columns before they could unite. Longstreet and his corps had not yet arrived. Bragg could not wait. He had to act now. Bragg chose the center column, that of Thomas, as his initial target. This would constitute the initiation of the series of engagements that became know as the battle of Chickamauga. Before we proceed, let us look at the cast of characters.

The Leading Characters

William Starke Rosecrans relieved Don Carlos Buell as commanding general of the Army of the Cumberland on October 30, 1862. He thus became the principal opponent of Confederate General Braxton Bragg for control of the state of Tennessee.

Strangely enough, Rosecrans had been a classmate of Buell at West Point in the class of 1842. However, their pre-war careers were quite different. Rosecrans graduated near the top of his class, fifth of 56, and entered the prestigious corps of engineers. Thus, he spent his entire 12-year pre-war active duty career in construction and teaching, and took no part in the Mexican War. He had absolutely no experience in troop command or fighting, having fought neither Mexican nor Indian.

He left the army in 1854 for civilian life, where he became an immediate success. He first took over a mining company, which prospered under his leadership, and then, with two partners, founded one of the first oil refineries, which was again a success. In addition, he held patents for several industrial inventions. It was clear that he was a man of great talent and, in all probability, would have become a man of prominence without the Civil War.

At the outset of the war, he was made a colonel by the governor of Ohio and quickly upped to general. He was McClellan's principal assistant in McClellan's successful and widely advertised campaign in West Virginia, and thus achieved national prominence almost from the beginning.

Rosecrans was subsequently transferred to the west where he gained further fame at Iuka and Corinth. At the time of his appointment as army commander, he was already a national figure. On the occasion of Rosecrans's assumption of command, a *New York Times* reporter wrote the following about him:

Map 26
Chickamauga — September 9, 1863

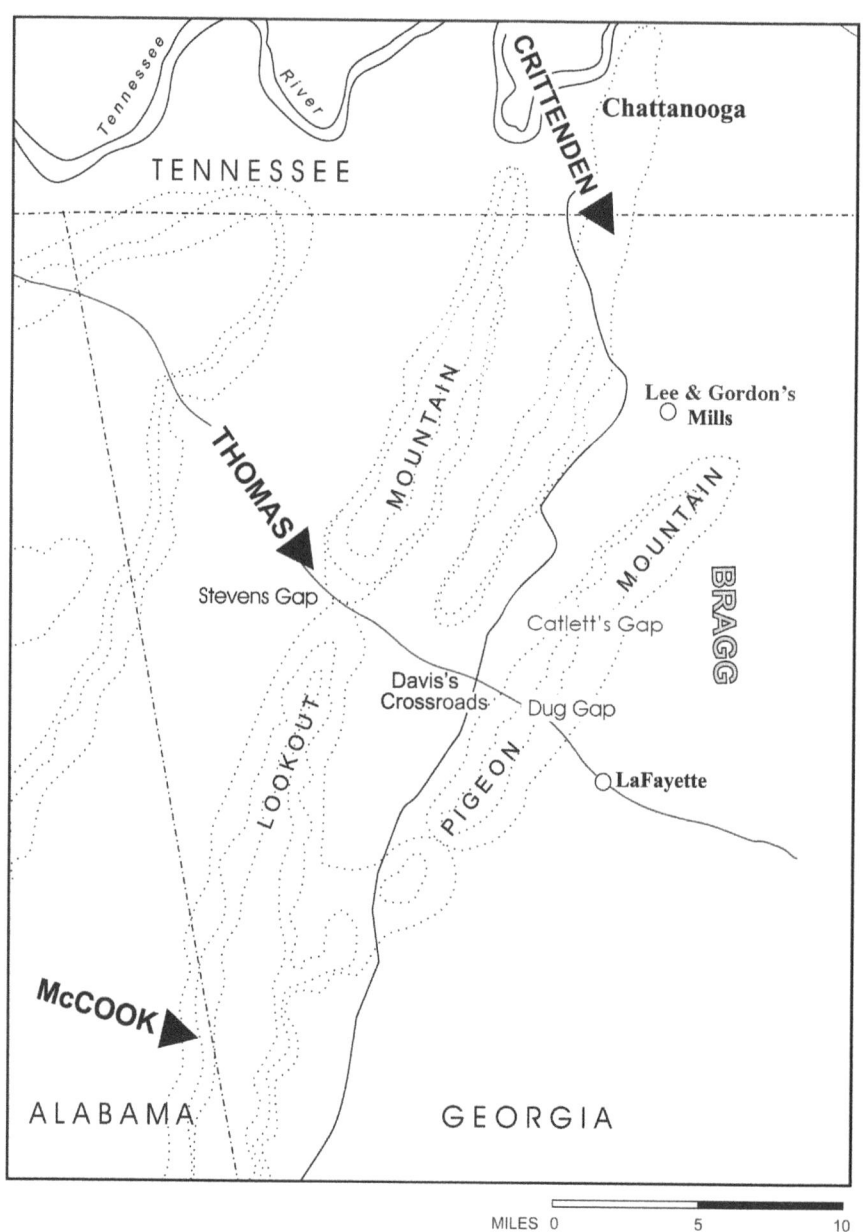

A change of commanders, and particularly if the change which has just been made, gives at least no dissatisfaction. Like Randolph, with his cup, we were beginning to say, "Anything for a change." Slow as our army was to lose confidence in and give up the leader to whose control it had so long been accustomed, it receives order No. 1 with hope fortified by the precedents of the new commander, that something more than marching and counter-marching will be done. Gen. Rosecrans has the confidence of the army and will prove himself worthy of it. His arduous, cautious and fatiguing campaign in Western Virginia, his subsequent career in Mississippi and his brilliant victory at Iuka entitle him to the confidence of the nation and the army. There is, at all events, more of pleasure than of pain at the change of leadership.[1]

Rosecrans did possess a certain charisma and was generally popular with his troops, but not to the degree of McClellan. His manner was described as brusque and his temper was short. He was a difficult subordinate and managed to alienate both Generals Halleck and Grant, and Secretary of War Stanton.

Be that as it may, his performance as army commander in its first year appeared to well justify his selection. With but a single major battle, that of Stone's River, by a series of brilliant maneuvers, he had succeeded in finessing Bragg out of most of the state. In September 1863, Rosecrans was at the peak of his popularity as he marched to his Waterloo at Chickamauga.

Now let us turn to the Confederate side. The quarrelsome Bragg was still commanding general of the Army of Tennessee. The amiable Bishop Polk, who often thought that he knew better than Bragg, was still second in command. However, the third in command, General Hardee, managed to get himself transferred out from under Bragg by a series of adroit, under the table, political moves. Hardee's replacement was General D. H. Hill from the east. Hill just happened to be a classmate, of the class of 1842, of General Rosecrans.

Hill was a brother-in-law of Confederate "Stonewall" Jackson and, like Jackson, an unusually religious man. However, Hill was noted for a distinctly unreligious characteristic. He was a notorious "croaker," that is, complainer. He constantly complained about the leadership, the system, etc. If there was one place a croaker was justified in croaking, it was in the command of Bragg; and Hill, instead of becoming a loyal subordinate of Bragg, was quickly drawn into the cabal of generals that was seeking Bragg's removal.

Bragg was to sum up Hill as follows:

> General Hill is despondent, dull, slow, and tho gallant personally, is always in a state of apprehension, and upon the most flimsy pretexts makes such reports of the enemy about him, as to keep up constant apprehension, and require constant re-enforcements. His open and constant croaking would demoralize any command in the world. He does not hesitate at all times and in all places to declare our cause lost.[2]

In addition to Bragg's regular command, he was, of course, to be augmented by various other units by rail in time for the climactic battle. One of these was commanded by General Longstreet, yet another member of the West Point class of 1842. Rosecrans, who graduated near the top of his class, was about to be given his comeuppance from a member who graduated near the bottom of his class, Longstreet.

The McLemore's Cove Operations

To pick up our story, by September 9 Bragg realized that he had a window of opportunity to destroy the Union columns in detail before they could unite. The Union columns of Crittenden in the north, Thomas in the center, and McCook in the south were too far separated one from another for mutual support. Bragg's army, on the other hand, was concentrated in the Lafayette area and located between the northern two Union columns. Bragg decided that the center of the Union columns, that of Thomas, would be his initial target, and that he would hit it as it entered McLemore's Cove.

To understand what transpired, we must stop here to take a closer look at the terrain involved (see map 27). Lookout Mountain was one of the major mountain ridges through which Thomas had to pass. As of the 9th, his lead division, that of Negley, was proceeding through Stevens Gap. However, once through Stevens Gap, he was still not clear of Lookout Mountain. Some miles to the south of Stevens Gap, a spur broke off Lookout Mountain, proceeded to the east, and then ran northward, parallel to Lookout Mountain. This spur was called Pigeon Roost Mountain. Unlike its parent, Pigeon Roost Mountain did not run all the way up to the Tennessee River, and thus formed a "U" with its parent, Lookout Mountain. The space within this "U" (i.e., the space between Lookout Mountain and Pigeon Roost Mountain) was called McLemore's Cove.

The road through Stevens Gap on which Thomas's men were proceeding ran eight miles eastward across McLemore's Cove until it hit Dug Gap in Pigeon Roost Mountain. Once through Dug Gap, the road continued on another four miles until it hit Lafayette, the central point of Bragg's concentrated army. A road ran north and south in the center of McLemore's Cove, and the point where it crossed the east-west road, the one leading from Stevens Gap to Dug Gap, was called Davis's Crossroads.

Bragg's intent as of September 9 was to destroy Negley's division and such other troops of Thomas as might join Negley during the 9th and 10th while they were passing through the cove. He proposed to do this by having Hindman's division proceed down the cove from its open-ended north on the north-south road, Cleburne's division of Hill's corps passing through Dug Gap on the east-west road, and the two meeting and hitting the enemy in the vicinity of Davis's Crossroads (see map 28). Bragg had cavalry troops under General Martin within the cove who, though not strong enough to challenge the enemy, could keep Bragg advised of their movements.

When we say that Bragg's army was concentrated at Lafayette, we really mean that Lafayette was near the center of the concentration. The army actually covered several square miles, and as a general statement, Polk's and Buckner's corps were to the north of Lafayette, and those of Hill and Walker were to the south. As of the night of the 9th, Bragg's headquarters were at Lee and Gordon's Mills, which was at the open end (the north end) of McLemore's Cove. It was here that Bragg issued his orders for the attack on the enemy forces in the cove. He issued separate orders to Generals Hill and Hindman as follows:

> Headquarters Army of Tennessee
> Lee and Gordon's Mills, Sept 9, 1863 — 11:45 P.M.
> Major General Hindman
> Commanding Division:
>
> General: You will move with your division immediately to Davis' Cross-Roads, on the road from Lafayette to Stevens' Gap. At this point you will put yourself in communication with the column of General Hill, ordered to move to the same point, and take command of the joint forces, or report to the officer commanding Hill's column according to rank. If in command you will move upon the enemy, reported to be 4000 or 5000 strong, encamped at the foot of Lookout Mountain at Stevens' Gap. Another column of the enemy is reported to be at Cooper's Gap; number not known.
>
> I am, general etc.
> Kinloch Falconer
> Assistant Adjutant-General[3]

The order to Hill was as follows:

11. Chickamauga

Headquarters Army of Tennessee
Lee and Gordon's Mills, Sept 9, 1863 — 11:45 P.M.
Lieutenant General Hill
Commanding Corps:

General: I inclose orders given to General Hindman. General Bragg directs that you send or take, as your judgment dictates, Cleburne's Division to unite with General Hindman at Davis' Cross-Roads to-morrow morning. Hindman starts at 12 o'clock to-night, and he has 13 miles to make. The commander of the column thus united will move upon the enemy encamped at the foot of Stevens' Gap, said to be 4,000 or 5,000. If unforeseen circumstances

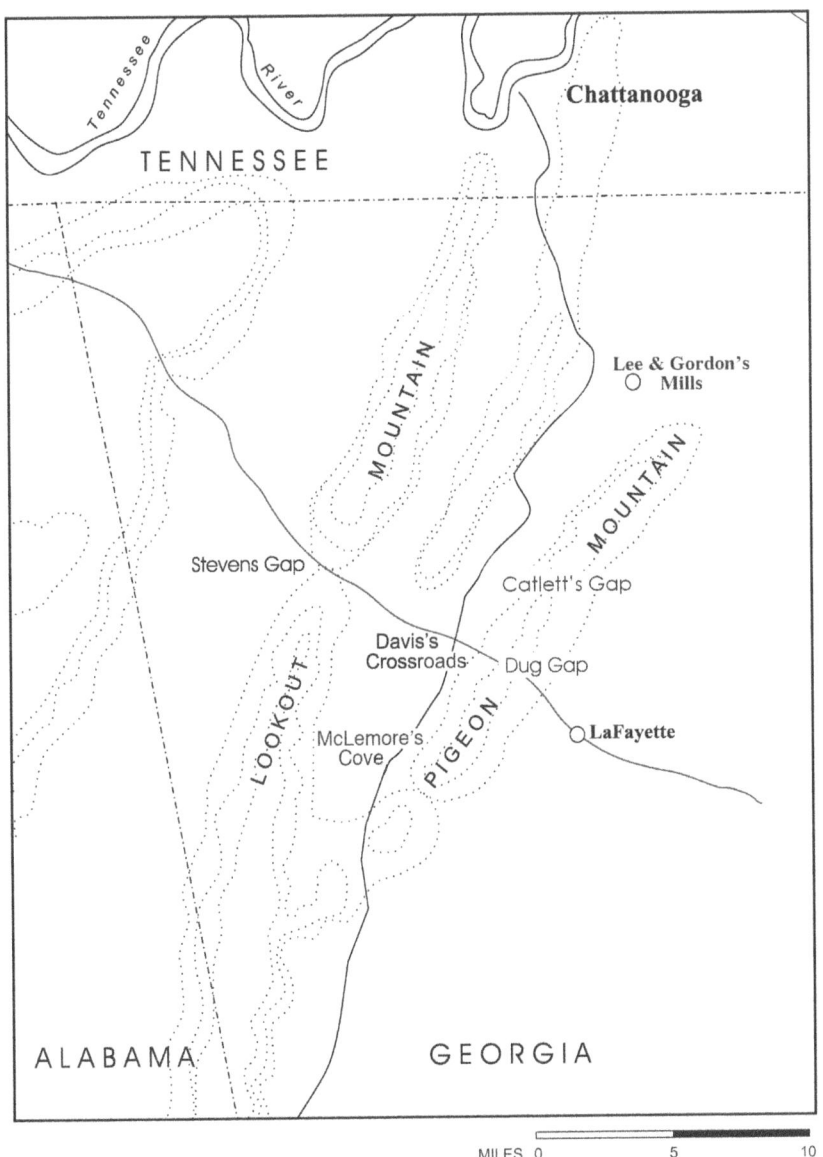

Map 27
Chickamauga — McLemore's Cove

should prevent your movement, notify Hindman. A cavalry force should accompany your column. Hindman has none. Open communication with Hindman with your cavalry in advance of the junction. He marches on the road from Dr. Anderson's to Davis' Cross-Roads.

I am, general etc.
Kinloch Falconer
Assistant Adjutant-General[4]

Map 28
Chickamauga — McLemore's Cove, September 9

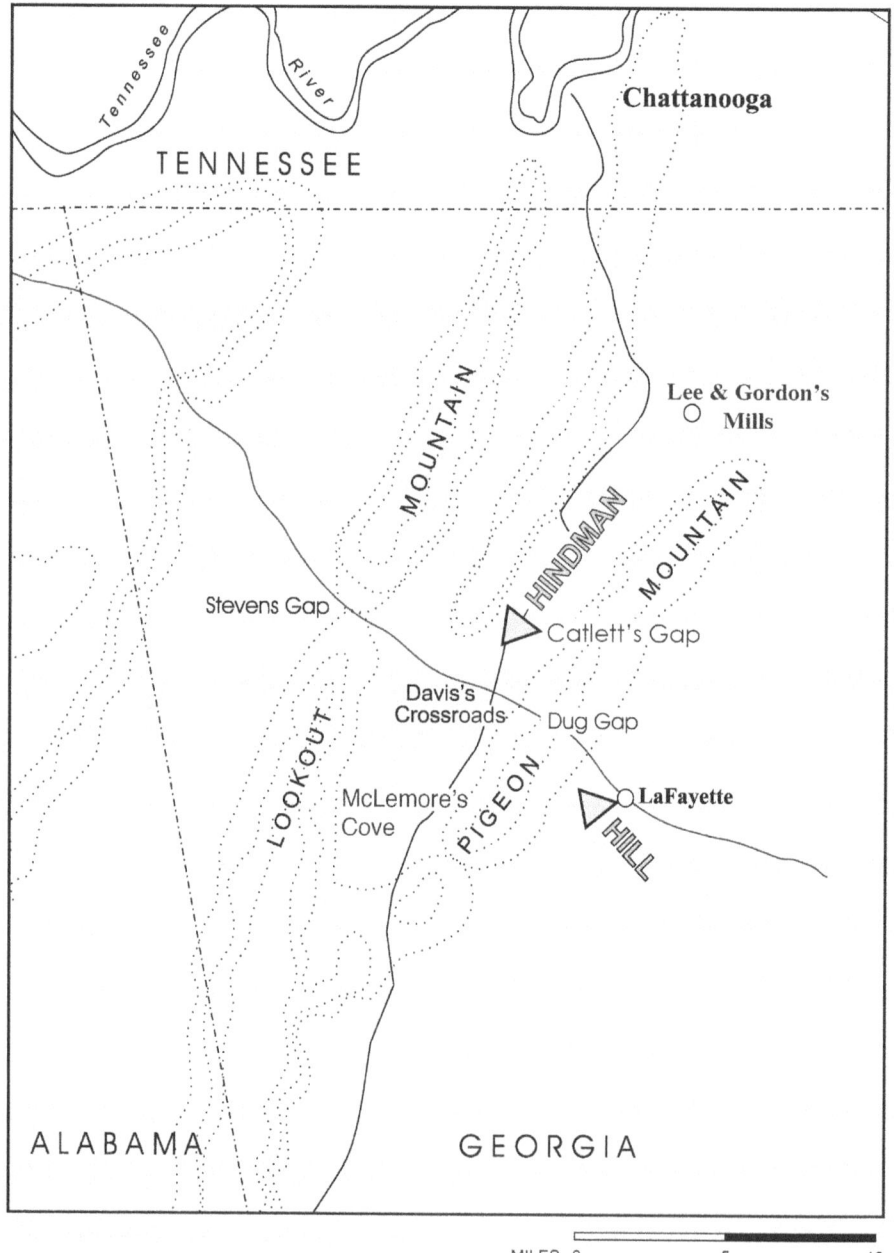

Hindman, whose command was located close to Bragg's headquarters, received his copy of the orders almost immediately, and quickly put his command on the road for a night march. By daylight, it had already reached Morgans, just four miles short of Davis's Crossroads, and he had still heard nothing from Hill. Hindman was reluctant to go further in the absence of any communication from Hill, and ordered the command stopped. He then drafted a message to Hill with a time of origin of 6 A.M. requesting information as to Hill's status. He sent it off by courier and then waited, and waited. His message did not even reach Hill until that afternoon.

In the meantime, Hill did not receive his order from Bragg until 5 A.M. Hill then decided that he could not implement Bragg's order. He prepared a message to Bragg so stating, listing a series of reasons. Some of the "reasons" were little short of frivolous, but one was at least partially valid.

General Martin, Bragg's cavalry commander within the cove, not knowing that Bragg would want to come into the cove via Dug Gap, had blocked both Dug and Catlett's Gaps with a plethora of felled trees so that the Union troops could not get out. Hill estimated that it would take 24 hours to clear the trees from the gap before he could enter. A more enterprising general than Hill might have resolved the problem and carried out the order and joined Hindman at Davis's Crossroads. Pigeon Roost Mountain was low and could be easily crossed by foot soldiers. The blocking of the gap only prevented the ingress of wheeled vehicles. Hill could have entered with his infantry while simultaneously clearing the gap for a later entry of his wheeled vehicles. But he did not, and Bragg accepted his reasons or excuses as they may be.

Bragg, however, upon hearing of Hill's inability to comply, was not to be denied. If he could not get Hill's men in through Dug Gap, he would send more men down the center road from the open-ended north. They would join Hindman, using the road that Hindman had used. He would use General Buckner's two divisions.

At 8 A.M., Bragg sent the following message to General Buckner, who was located nearby, by courier:

Headquarters Army of Tennessee
Lee and Gordon's Mills, Sept 10, 1863 — 8 A.M.
Major General Buckner
(at) Anderson's:
   General: I enclose orders issued to Generals Hill and Hindman. General Hill has found it impossible to carry out the part assigned to Cleburne's division. The general commanding desires that you will execute without delay the order issued to General Hill. You can move to Davis' Cross-Roads by the direct road from your present position at Anderson's along which General Hindman has passed.
   I am, general etc.
   George Wm Brent
   Assistant Adjutant-General[5]

Both Hindman and Hill were notified of Buckner's orders.

So, we had the following situation during the day of the 10th. General Hindman was stopped at Morgan, on the central cove road, four miles short of Davis's Crossroads. General Buckner, with two divisions, was marching down the central cove road to join Hindman. General Cleburne's troops were unable to enter the cove via Dug Gap because it was blocked with felled trees, but were busily engaged in clearing the trees. While all of this Confederate activity was taking place, the Union forces were not idle. A second division of General

Thomas's central force was passing through Stevens Gap to join the division of General Negley already in the cove, thus raising the number of Union troops in the cove to about 10,000. This was not bad news for the Confederates. If all went well, it should merely increase the size of the bag of prisoners.

The head of General Buckner's column reached Hindman at Morgans at about 4:45 P.M. and Buckner reported to Hindman, who was senior for orders. Hindman decided that it was now too late to make an attack on the Union forces, and ordered all three divisions into bivouac. Thus, we came to an end of the activity of September 10. The first day of Bragg's window of opportunity had been wasted.

If Bragg's window of opportunity was wasted on the 10th, he did not intend to waste the 11th. Throughout the evening and night, he continued to receive alarming reports that the three Union columns were moving to converge on Lafayette. He intended to exert every effort to conclude the operations in the cove on the 11th. To this end, he took a number of actions on the night of September 10–11. He ordered that Dug Gap be cleared by daylight. He ordered Walker's reserve corps to join Cleburne's division at Dug Gap and be prepared to join Cleburne in an attack on the Union troops in the cove as soon as they heard Hindman's guns in the morning. He moved his headquarters in the night from Lee and Gordon's Mills to Lafayette so as to be in a better position to control events in the cove on the morrow. Lastly, he sent several messages by courier during the night to Hindman.

Bragg sent messages by courier to Hindman with times of origin of 6 P.M. and 7:30 P.M. urging him to be expeditious. The 6 P.M. message was:

> Headquarters Army of Tennessee
> Gordon's Mills, Sept 10, 1863 — 6 P.M.
> Major General Hindman
> Commanding Division
>
> General: The general commanding instructs me to say that Crittenden's corps marched from Chattanooga this morning in this direction and that it is highly important that you should finish the movement now going on as rapidly as possible.
>
> I am, very respectfully, your obedient servant.
> George Wm Brent
> Assistant Adjutant-General[6]

The 7:30 P.M. message was:

> Headquarters Army of Tennessee
> Gordon's Mills, Sept 10, 1863 — 7:30 P.M.
> Major General Hindman
> Commanding etc.

Major General T. C. Hindman CSA: Always competent except in the "cove."

General: The enemy is now divided. Our force at or near Lafayette is superior to the enemy. It is important now to move vigorously and crush him.
I am, very respectfully, your obedient servant.
Kinoch Falconer
Assistant Adjutant-General[7]

While Bragg was sending these messages to Hindman, Hindman, Buckner and the other generals of their command in the cove were having a meeting to discuss their situation. They were apprehensive and fearful that they were heading into a trap. They feared that the force in front of them was growing, their potential exits out of the cove were blocked by fallen trees, and that Crittenden might appear in their rear. They suggested an alternative course of action. Instead of continuing down the cove, why not join in an all-out attack on Crittenden to their north. Hindman dispatched Major Nocquet of Buckner's staff to lay this proposal before Bragg.

Nocquet finally tracked down Bragg at Bragg's new headquarters at Lafayette at midnight and laid down the proposal before Bragg. Bragg heard him out and then said no. Hindman must adhere to the original plan. Bragg then sent his final message of the night to Hindman. It was:

Headquarters Army of Tennessee
Lafayette Ga. Sept. 10, 1863 — 12 P.M.
Major General Hindman
Commanding etc.

General: Headquarters are here, and the following is for information: Crittenden's corps is advancing on us from Chattanooga. A large force from the south has advanced to within 7 miles of this point. Polk is left at Andersons to cover your rear. General Bragg orders you to attack and force your way through the enemy to this point at the earliest hour that you can see him in the morning. Cleburne will attack in the front the moment your guns are heard.
I am, general etc.
George Wm Brent
Assistant Adjutant-General[8]

When the sun came up on the morning of September 11, 1863, to begin another beautiful late summer day, Bragg was loaded for bear. He had 30,000 men poised to jump 10,000 Federals in the cove. This time there were no blocked passes, and no long marches were required. Everyone was in place and close by. How could he lose? It would all kick off as soon as Cleburne heard Hindman's guns.

As the sun rose higher and higher, there was still no sound of guns, but the full panorama of the deployed Confederate forces became ever more visible to the Union observation posts on Lookout Mountain. By 10 A.M. there was still no sound of Hindman's guns, but every regiment of the Confederate forces was now clearly visible to the observation posts. The Union scouts observed, they reported, and the Union generals discussed.

By noon there was still no sound of Hindman's guns, and the Union generals had decided that they had better get their troops out of the cove. Finally, at mid afternoon, the guns sounded and the Confederates advanced, just as the last Union troops were disappearing through Stevens Gap. The Confederates had blown their last opportunity to win a smashing victory in the cove during Bragg's "window of opportunity."

What went wrong? Bragg's intention was to destroy the Union forces in McLemore's Cove during his window of opportunity when Thomas could not be supported by either

Crittenden in the north or McCook in the south. His orders to Hill and Hindman never specifically tasked them with doing just that. His last order to Hindman, issued at midnight on September 10–11, was particularly damning. We know that Hindman, rather than worrying about destroying the Union forces in the cove, was concerned about his own troops' safety. He expressed this view in sending Major Nocquet to Bragg during the night of September 10–11. Hindman was afraid that he would be boxed in by the forces in his front, and Crittenden in his back, and that he was inhibited from getting out the side by the blocked passes. We can reasonably conclude that Hindman would act quite differently if he thought that his task was to save his troops from destruction, rather than to destroy the enemy.

Bragg's order of midnight to Hindman, after conferring with Major Nocquet, contained the phrase "General Bragg orders you to attack and force your way through the enemy to this point at the earliest hour...." This confirmed in Hindman's mind that, indeed, Bragg thought him trapped, and his main task was to get out. Here we will turn to Hindman's own words:

> My construction of the above-quoted dispatch was that the general commanding considered my position a perilous one, and therefore expected me not to capture the enemy, but to prevent the capture of my own troops, forcing my way through to Lafayette, and thus saving my command and enabling him to resist the forces that seemed about to envelop him. This idea only was conveyed by the language used. Keeping it in view, I delayed issuing the order of march until the scouting parties sent toward Lookout Mountain should report, and in the hope, also, of hearing from army headquarters and from General Hill in answer to the important letters sent by Major Nocquet, or the one of 9:10 P.M. of the 10th sent by courier.[9]

Hindman later offered a variety of reasons (or excuses) for his failure to attack. There is little doubt, however, that a primary reason was a misunderstanding of his mission, which was confirmed in his mind by the unfortunate wording of Bragg's midnight order.

Two days of Bragg's window were now wasted. Although he had not defeated Thomas in the center, he had at least momentarily pushed him back and Crittenden was still alone to his north.

At the end of the 12th, Bragg had lost his chance for a one-sided victory in McLemore's Cove, but his window of opportunity had not yet quite closed. The three Federal corps were converging but still some distance apart, and his main force was still between those of Thomas and Crittenden. That night Bragg got a report that Crittenden was moving across his front and that Polk was positioned to hit Crittenden's lead division in isolation.

One of the principles of good battlefield order writing was to measure the degree of discretion the order allowed to the nature of the general receiving the order. Some generals would do their best to carry out the spirit of the order even though the order may not have coincided with their own views. Others would attempt to substitute their own judgment to the degree the order allowed. The subordinate to whom Bragg would have to turn to for the attack on Crittenden's division was Bishop Polk, and Polk definitely belonged to the latter category. He subconsciously considered that his judgment was better than Bragg's. Any orders Bragg would have to send to Polk must contain as little "wiggle room" as possible if Bragg expected the order to be carried out.

With this in mind, let us look at Bragg's order to Polk:

Headquarters Army of Tennessee
Lafayette Ga. Sept. 12, 1863 — 6 P.M.
Lieutenant-General Polk:

General: I enclose you a dispatch from General Pegram. This presents you a fine opportunity of striking Crittenden in detail, and I hope you will avail yourself of it at daylight tomorrow. This division crushed, and the others are yours. We can then turn again on the force in the cove. Wheeler's cavalry will move on Wilde, so as to cover your right. I shall be delighted to hear of your success.

Braxton Bragg[10]

Polk responded with a lengthy analysis of his situation, why he expected to be attacked in the morning, why he needed more men, and why he would have to take a defensive posture.[11] Bragg responded with a message indicating that Polk would be well reinforced, but ordered him not to defer his attack because the force he had was already numerically superior to the enemy.[12] In the event, Polk never made the attack, and Bragg's final opportunity to destroy the enemy in detail was lost. Polk's performance was shades of Perryville. In just one more transaction, Bragg decided he must be rid of Polk.

Bragg had wasted his window of opportunity by bad order writing and bad generalship on the part of Hill, Polk, and Hindman. The final two days of the battle were to take place on September 19 and 20, 1863. By this time, the Federals had united in the area of Lee and Gordon's Mills. Bragg received his final reinforcements, those of Longstreet, on the night of September 19. At this point, he had a substantial numerical superiority over Rosecrans, and he hoped to do on the 20th, by brute force, what he had failed to accomplish by finesse up to that point — that is, the destruction of Rosecrans's army.

At the last minute, he reorganized his army into two wings. Longstreet would command on the left and Polk on the right. September 20 would be the decisive day. Polk would open the battle by attacking at dawn, and Longstreet would attack when Polk's attack was having its maximum effect.

Things started out badly for the Confederates because of bad coordination and order dissemination during the night. Polk's "dawn" attack did not get off until between 9 and 10 A.M. At this point, one might conclude that the inept Confederates were going to blow their final chance. However, there was to be one more instance of bad battlefield order writing during the battle, and this instance was to be the granddaddy of them all. In this case, the bad battlefield order writing would not be at the hands of the Confederates, but of the Union.

Polk's attack was against Thomas's side of the Union line, and the severely stressed Thomas kept calling to Rosecrans for reinforcements. Rosecrans, in turn, kept transferring men from the inactive side of his line to Thomas. At midmorning, the sequence of units in Rosecrans's line was as indicated on figure 1. From this point on, we run into controversy as to what transpired and whose fault it was. This controversy extends to this day.

General Rosecrans, in a letter to the War Department dated January 13, 1864, alleged that sometime during the morning of the 20th, exact time unspecified, General Brannan was given an order to leave his position and support the left (i.e., Thomas). However, Brannan, finding his skirmishers already engaged with the oncoming troops of Polk, requested to know if he should move under the circumstances. Brannan was told to stay where he was. Thus, according to Rosecrans, Brannan was first told to move, and then to stay.[13] There is no record of any such orders to Brannan, and neither Brannan nor Thomas nor Rosecrans

Figure 1.
Chickamauga — The Fateful Break in the Union Line.

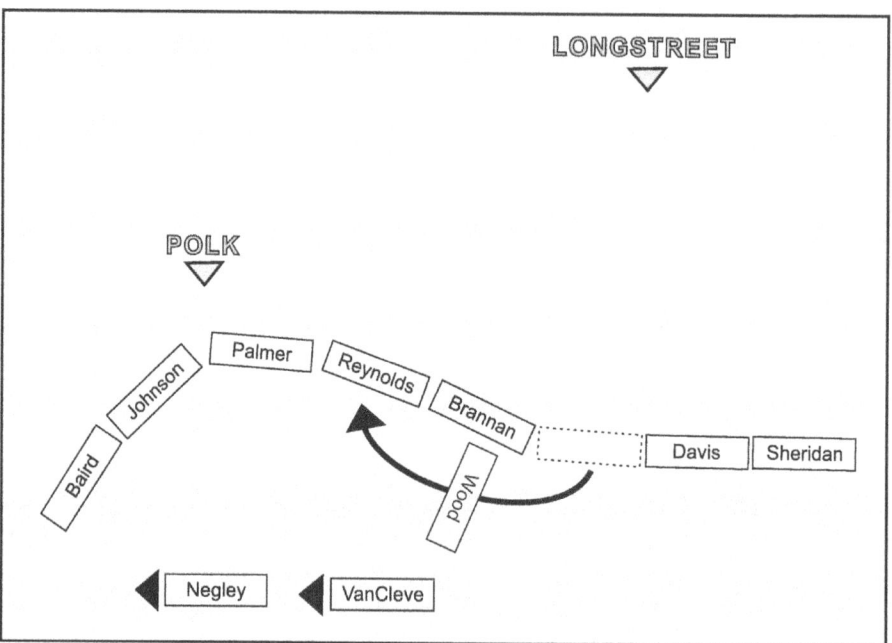

mentioned such a transaction in their official report of the battle submitted shortly after the event. In any event, if such orders to Brannan had taken place, it is indisputable that General Wood, who was immediately to Brannan's right, was not notified of them.

Sometime later, three couriers from Thomas came to Rosecrans in quick succession asking for reinforcements. The first was directed to say that General Negley had already gone and should be near at hand, and that Brannan's reserve brigade was available. The second was directed to say that General Vancleve would at once be sent to his assistance, which was accordingly done. The third to arrive was Captain Kelly, Thomas's aide-de-camp. Kelly announced that Thomas was being heavily pressed and that General Brannan was out of line and that General Reynolds's right was exposed.[14]

It was at this point that the infamous battlefield order came into existence. The order was called for by Rosecrans and drafted by his aide-de-camp, Major Bond. It was as follows:

Headquarters, Sept 20—10:45 A.M.
Brig General Wood, Commanding Division etc.:
  The General Commanding directs that you close upon Reynolds as fast as possible, and support him.
  Frank S. Bond
  Major an A. D. C.[15]

Let us now see what General Wood knew or did not know when he received the order. First of all, his division was deployed in a thickly wooded area, and from a central position he could not see either end of his own line. The only instructions he had received that morning were to ensure that his left abutted the right of Brannan. To this end, he placed a staff officer at the junction, whose only function was to see that the two were connected.

At the time Wood was handed the message, they were. Wood knew nothing of the alleged orders to move and not move Brannan that supposedly took place earlier in the day. He did not know anything of the line of battle other than that Brannan was to his left and Davis to his right. He did not know where Reynolds, whom he was ordered to support, was located in the line. He knew that the battle was raging to his left and could see that troops were constantly being sent from his right to his left. Consequently, he rightly concluded that Reynolds, whom he was ordered to support, must be somewhere to his left.

When the order was handed to Wood, the envelope was marked "Gallop." This was an instruction to the courier. Horseflesh was important in the Civil War and horses were not to be wasted. Consequently, couriers were given directives as to the rate they were to proceed. "Gallop" indicated urgency. It was equivalent to a priority. The message said "close upon Reynolds as fast as possible." This, plus "Gallop," indicated to Wood that Reynolds was in trouble and he must act as quickly as possible. At the time he received the message, he was not engaged in his front and had no indications that an attack on his front was imminent. As to creating an opening in the line because of his withdrawal, he considered that Rosecrans knew what he was doing and would provide for the closing of the gap if he thought necessary.

Thus, Wood did what any competent general would do under the circumstances. He pulled his division out of line, and moved along behind Brannan's line of battle to his left, toward the battle, until such time as he would encounter Reynolds, whom he was ordered to support. "Support" normally meant to take a position in the rear of the one supported.

At the very time Wood was withdrawing, Longstreet, just out of sight, was preparing his attack. When it came, it came like a thunderbolt, hit thin air, and passed through the break in the line left by Wood. Longstreet was now on the flanks and rear of the Union line, and it disintegrated, except for Thomas, and fled in panic. This was the first rout of a Union army since first Bull Run, and had Bragg properly followed up his success, the war might have been turned around in favor of the Confederates that very afternoon.

This was a Union disaster, and Rosecrans blamed Wood. Rosecrans's accusations against Wood were twofold. First, Rosecrans said that if Wood, having been told to support Reynolds, who was beyond Brannan, found that Brannan was still there, he should have asked Rosecrans, who was nearby, for clarification. Nonsense! If Rosecrans was so nearby, why did he not notice Wood removing 5,000 men from the line? If he did notice, why did he not say something?

Secondly, Rosecrans said that if Wood was already under attack by Longstreet, he should not have moved. Wood demonstrated conclusively by affidavits of those present that no such attack was in progress when he withdrew from the line.

The cause of the Union disaster was entirely due to Rosecrans's faulty battlefield order to Wood. Rosecrans was obviously confused by the orders to Brannan to go, to stay, etc. His order to Wood should have adhered to the principle of "what" do you want him to accomplish, not "how!" The order should have read:

> Ensure you are closed up to the command to your left. The line must remain unbroken. Davis has been given similar instructions to keep closed to your right.

Had that been the wording of the order, the disaster would never have happened.

12

Spring Hill

In this chapter we will address what many historians would consider the most egregious foul-up of the war. This incident had such repercussions that it significantly affected the final outcome of the war. The foul-up was at the hands of the Confederates and occurred during the battle of Spring Hill on November 29 and 30 of 1864. The battle has been studied and restudied, and to this day, there is no consensus among historians as to just who was responsible for the debacle. Our subject is battlefield order writing and, admittedly, battlefield order writing was not the sole cause or even the main cause. However, it did play a significant role and we will concentrate on this aspect. We must start at the beginning.

The Situation Leading Up to Spring Hill

In the summer of 1864, the two main theaters of war were as they were at the beginning: Virginia and Tennessee. It was here that the main armies contended—Lee against Grant in the east, and Johnston against Sherman in the west.

In the western theater, Johnston continued to retreat before Sherman, trading space for casualties until the theater of war advanced from Tennessee into Georgia and the key city of Atlanta was endangered.

Confederate President Davis, who never liked Johnston in the first place, became ever more exasperated with Johnston's retreats, and finally relieved him of command on July 17, 1864. Davis's choice for his replacement was the young John Bell Hood. This may have been Davis's biggest mistake of his presidency, but more about this later. Davis wanted fighting and not retreat, and Hood was certainly the man for that.

Despite the change of command, Atlanta was lost to Sherman on September 3. At this juncture a strange thing happened. The two contending armies separated. Hood realized that he could not evict Sherman from Atlanta and decided that he would attack Sherman's long and vulnerable supply line. Sherman would then be forced to evacuate Atlanta and move north. Sherman, however, had other ideas. He decided to burn Atlanta, completely relinquish his supply line to the north, and march through Georgia to the sea, living off the land. Thus began Sherman's famous "march to the sea."

Hood, thus finding himself with no supply line to cut, had to devise a new strategy. He decided he would reconquer Tennessee and march into Kentucky. Then, no matter how far Sherman had penetrated into Georgia, he would have to return to confront Hood.

When Sherman started his march, he considered the possibility that Hood would

12. Spring Hill

attempt the reconquest of Tennessee. Consequently, he left behind a force he considered sufficient for its defense. This force consisted of two separate contingents: a force of about 19,000 under General Thomas based at Nashville to hold west Tennessee, and a force of about 20,000 centered at Chattanooga under General Schofield to hold east Tennessee. Thomas was the senior of the two and thus responsible for the overall defense of Tennessee.

Hood commanded a force of some 30,000. Thus, he was superior to either of the two Union defense forces, but inferior if they combined. In order for his plan to succeed, he must destroy them separately.

By late fall, it was apparent that Hood was heading for Nashville. It was thus imperative that Schofield's force get there before Hood. Conversely, Hood must cut off Schofield before he could get to Nashville. It is here that our story of the battle of Spring Hill really begins — the race for Nashville.

The first point where Hood hoped to cut off Schofield from Nashville was at Columbia, Tennessee, which was on the pike to Nashville. In order to understand what subsequently transpired, we must examine the pike from Columbia to Nashville, so let us take a stroll up the pike (see map 29). The pike ran roughly north by northeast. Columbia was on the south side of the Duck River, and a bridge crossed the Duck at Columbia. The Duck was a substantial river and could only be crossed by wagons at a bridge or occasional ford. Proceeding up the pike toward Nashville, we come to Rutherford's Creek, seven and a half miles from Columbia. Rutherford's Creek had no bridge and was no more than an inconvenience to cross. Two and a half miles beyond Rutherford's Creek, we come to Spring Hill. Spring Hill was merely a village on a crossroads. Spring Hill was thus ten miles up the pike from Columbia. Three miles beyond Spring Hill, we come to a railroad crossing named Thompson's Station. Eight miles farther up the pike toward Nashville, we come to the town of Franklin. Franklin was on the south side of the Harpeth River. The Harpeth was substantial and could only be crossed by bridge or occasional ford. Both a vehicular and railroad bridge crossed the Harpeth at Franklin. Franklin was thus 21 miles up the pike from Columbia and 11 miles up the pike from Spring Hill. Another 15 miles up the pike from Franklin, one came to Nashville. Nashville was about 36 miles up the pike from Columbia.

When Schofield reached Columbia a hair's breadth ahead of Hood on November 28, 1864, he was nearing safety. He had been told that reinforcements from Missouri would be awaiting him at Franklin, and he was now on the pike just 20 miles short of Franklin.

It was here at Columbia that Hood was to make his finest move. He divided his force. He left two division of S. D. Lee's corps and most of his artillery on the pike south of Schofield to conduct demonstrations as if he were about to attack Schofield. He then took the remainder of his command, that is, the corps of Cheatham and Stewart and the division of Johnson from Lee's corps, five miles down the river where he crossed by a ford. He then made a forced march via back roads to Spring Hill on the pike in Schofield's rear.

Meanwhile, Schofield, thinking that he was about to be attacked, ordered his 800-wagon supply train, guarded by a 5,000-man division under General Stanley, to proceed up the pike to Spring Hill to presumed safety from the oncoming battle. The wagon train, accompanied by Stanley's troops, reached Spring Hill between 2 and 3 in the afternoon on the 29th. Within an hour or so of the wagon train's arrival, Hood's flanking force, headed by General Cheatham's corps, began arriving at Spring Hill.

At this point, if we were to consider that Hood and Schofield were engaged in a draw poker game, Hood had been dealt four aces on the initial deal. He had not only cut off Schofield from safety at Franklin and Nashville, but he had cut him off from his own 800-wagon supply train.

They were in a war-ravaged part of Tennessee where Schofield could not subsist off the land, and without his supply train he would starve. Within an hour, Hood's arriving troops would have four times as many troops in Spring Hill as Stanley, and so the defeat of Stanley and the seizure of the supply train should have been a slam dunk. Schofield's

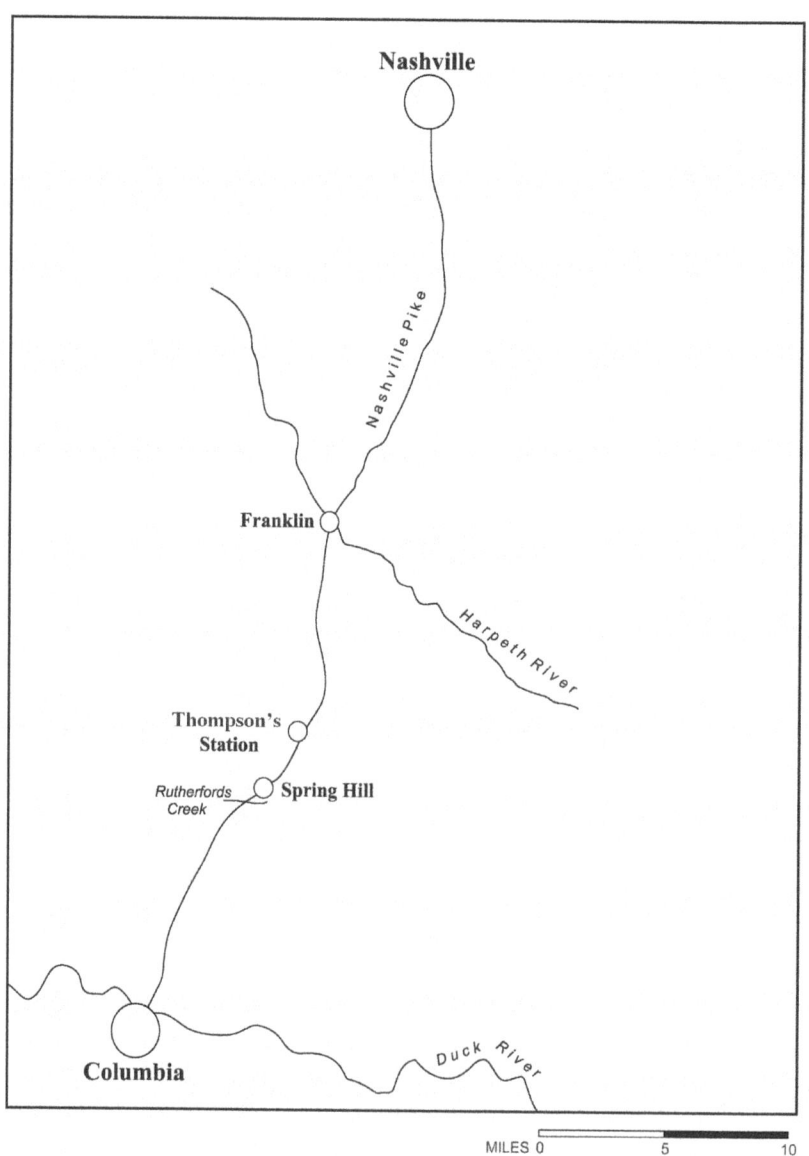

Map 29
Spring Hill

General John B. Hood CSA: An invalid in over his head.

Major General J. M. Schofield USV: A winner before, during and after the war.

main force was still ten miles south on the pike at Columbia, frozen in place by an expected attack by the troops of S. D. Lee that Hood had left behind.

Poor Stanley was in an unenviable position. He had no time to dig in prior to the anticipated attack and he could not even choose the ground for defense. He had to maintain a position between the wagon train and the enemy. The stage was set. The battle of Spring Hill was about to begin.

But before we describe the battle of Spring Hill, let us take a brief look at the two main players in our drama, Generals Hood and Schofield.

The Main Characters

Our two leading characters were John Bell Hood and John McAllister Schofield. These were two young men who were famous nationwide in 1864 thanks to the Civil War. Had it not been for the war, as of 1864, both would have been obscure lieutenants in the regular army.

The two had much in common. As of 1864, each was 33 years old and each was his nation's youngest army commander. Both had graduated from West Point in the class of 1853. Schofield was the more intellectual of the two, having graduated near the top of his class (seventh of 52), while Hood finished near the bottom (44th of 52). Furthermore, Hood narrowly avoided expulsion because of excessive demerits.

Both quickly rose in rank to general and both distinguished themselves in combat from the early days of the war. Schofield received the Medal of Honor for gallantry in leading a charge at Wilson's Creek, and Hood won distinction in almost every battle in the east from the peninsula to Gettysburg. He was usually considered Lee's best division commander.

By the fall of 1864, the futures of Hood and Schofield began to diverge. Hood had been seriously wounded twice and was now a physical wreck. At the battle of Gettysburg, he received a serious wound in his left arm that resulted in it being useless for the rest of his life. At Chickamauga, he received a serious wound in his right leg that resulted in it being amputated just below the hip. He was now an invalid and had to walk with crutches. He could not mount his horse alone and was frequently in pain. It was arguable if he should be kept on active duty.

It was in this condition that President Davis considered Hood a potential replacement for General Johnston as commanding general of the second biggest army in the Confederacy. Davis consulted Lee. Lee's reply was ambiguous and lukewarm. Lee wrote:

> It is a bad time to release the commander of an army situated as that of Tenn. We may lose Atlanta and the army too. Hood is a bold fighter. I am doubtful as to the other qualities necessary ... Hood is a good fighter, very industrious on the battlefield, careless off and I have had no opportunity of judging of his action when the whole responsibility rested on him.[1]

Nevertheless, Davis appointed Hood army commander. In connection with this, he was elevated to four stars, albeit temporarily. This was a distinction achieved by only six other Confederates. In the five months between July 18, the date of Hood's appointment, and the battle of Nashville on December 15–16, 1864, Hood led the Army of Tennessee to destruction. In retrospect, he was clearly appointed to a level of command beyond his capabilities. Hood was relieved of command of the Army of Tennessee and spent the remainder of the war in relatively unimportant positions.

Hood's post-war career was a failure. He entered the insurance business, which failed, and he died prematurely at the age of 48 in poverty.

While Hood's career waned, that of Schofield waxed. He remained in top command and in good favor until the end of the war. In June 1868, he was appointed Secretary of War by President Johnson, a distinction that only 27 other Americans had achieved to that date. In 1888, he was appointed to the top post in the army, that of Commanding General of the United States Army. He remained in this capacity until his retirement seven years later. In the eyes of many, he was the most effective peacetime commanding general of all time. To this date, West Point cadets are required to memorize a portion of a Schofield speech as follows:

> The discipline which makes the soldiers of a free country reliable in battle is not to be gained by harsh or tyrannical treatment. On the contrary, such treatment is far more likely to destroy than make an army. It is possible to impart instruction and give commands in such a manner and such a tone of voice as to inspire in the soldier no feeling, but an intense desire to obey; while the opposite manner and tone of voice cannot fail to excite strong resentment and a desire to disobey. The one mode or the other in dealing with subordinates springs from a corresponding spirit in the breast of the commander. He who feels the respect due to others cannot fail to inspire in them the respect for himself. While he who feels and hence manifests, disrespect for others, especially his subordinates, cannot fail to inspire hatred for himself.[2]

The names of Schofield and Hood live on to this day: Schofield in "Schofield Barracks," Hawaii, and Hood in "Fort Hood," Texas.

The Battle of Spring Hill

General Stanley's division with the wagon train no sooner arrived at Spring Hill than it was attacked by Forrest's cavalry. Forrest, however, was not strong enough to break through to the wagon park and was repulsed. As Forrest was being repulsed, Hood's flanking force, headed by Cheatham's corps, began to arrive.

Hood ordered Cheatham to attack immediately. The attack was delivered by the first of Cheatham's divisions to arrive on the scene, that of his most competent subordinate, Cleburne. The attack was vigorous and almost succeeded, but not quite. It was now between 4:30 and 5:00 P.M. It would not be dark until 7 P.M. There was still time for another attack. If an attack by one of Cheatham's three divisions almost succeeded, an attack by all three, or even any two of the three, ought to be a slam dunk. Cheatham's divisions of Bate and Brown were arriving and Cheatham was busily assigning them their attack positions. There were some problems in alignment and it was getting dark. Hood finally decided that it was too dark and postponed the attack until daylight on the 30th.

Hood was dissatisfied with Cheatham's performance. He was later to write:

> Major General Cheatham was ordered to attack the enemy at once vigorously and get possession of the pike. And although these orders were frequently and earnestly repeated, he made but a feeble and partial attack, failing to reach the point indicated. Had my instructions been carried out, there is no doubt we should have possessed ourselves of the road.[3]

The Night of November 29–30, 1864

As darkness fell on the 29th, Hood was disappointed but not discouraged. He still thought he held a winning hand. After all, as he saw it, his main force was between Schofield's main force at Columbia and Schofield's wagon train and Stanley's troops at Spring Hill. Stanley could not escape northward toward Franklin during the night as Forrest's cavalry now held Thompson's Station on the pike to the north. Cheatham would destroy Stanley and seize the wagon train. Schofield's main force could not escape to the north (or so he thought) because of his intervening flanking force, and it certainly could not escape to the south. Lee was to the south of him and, furthermore, Schofield's force had moved to the north side of the Duck River for defense. Now the river would be a barrier for any attack to the south. Furthermore, moving south would only ensure Schofield's starvation. He would be in an impoverished area sans his supply train. As Hood saw it, it was all over but the celebration.

When Hood moved north to Spring Hill earlier on the 29th with his flanking force of Cheatham's and Stewart's corps, he sent Cheatham on ahead to attack at Spring Hill while he stopped Stewart's corps at Rutherford's Creek, two and a half miles short of Spring Hill, until he decided where to commit it. Now, after dark, he returned to Stewart. He ordered Stewart to proceed north beyond Cheatham's position and to take a blocking position astride the pike facing south. Thus, Stewart would have a dual purpose: he could assist Cheatham with the destruction of Stanley at daylight, and he would ensure no escape of any of Schofield's troop northward toward Franklin. It was now dark and Hood provided Stewart with a guide to show him his new position. Stewart set out with the guide into the confusing darkness. Before they succeeded in locating Stewart's new intended position, a

**Map 30
Spring Hill — Bivouac Areas, Night of November 29, 1864**

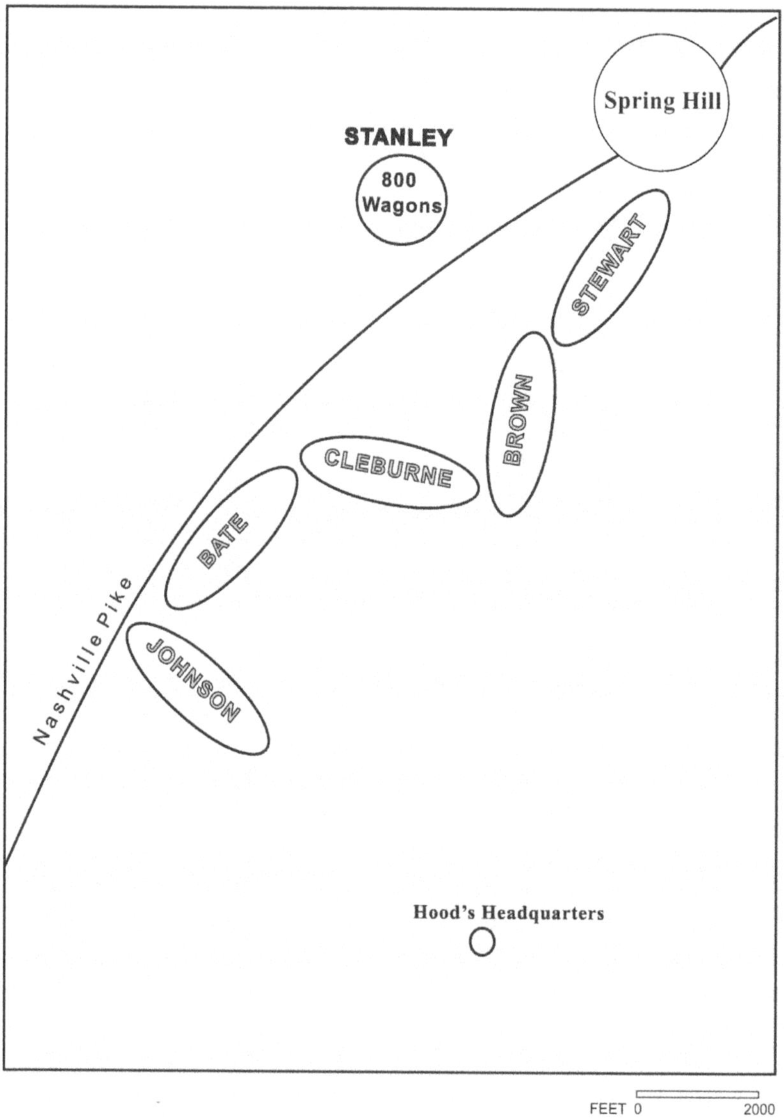

courier reached them. This was an officer from Cheatham's staff purportedly carrying new orders for Stewart from Hood. The new orders called for Stewart to form on Cheatham's right. At this point, things just did not seem right to Stewart. First, why would an officer from Cheatham's staff bring him an order from Hood? Second, and more important, the implementation of this new order was diametrically opposed to the one he had just received from Hood. Instead of placing his corps across the pike, it would angle the corps away from the pike. At this point, Stewart decided to put his corps, tired of marching and counter marching, into bivouac and go see Hood himself. The place he put his troops in bivouac was on the same side of the pike on which Cheatham's troops were in bivouac. No one was astride the pike (see map 30 for bivouac positions).

It was getting late when Stewart arrived to see the tired Hood. Hood said, yes, he had sent the second order, that after he left Stewart, Cheatham came to see him, still concerned about the alignment of his troops for the intended dawn attack on Stanley. Cheatham assured him that this placement of Stewart would cover the pike. In any event, everything was all right, tomorrow would bring victory, and Stewart should let his troops rest. And so matters came to rest that night for the Confederates: Stewart and Cheatham camped on the same side of the pike.

While the Confederates bedded down for the night, Schofield was busier than a one-armed paper hanger. By dark that night of the 29th, that is, 7 P.M., Schofield finally decided that his troops at Columbia were not going to be attacked and ordered them to march up the pike to Spring Hill. The order of march was to be Cox's division of the Twenty-Third Corps; the third division of the Fourth Corps; and then, bringing up the rear, the first division of the Fourth Corps. The march was to start at 7 P.M. Skirmishers were to be left behind facing Lee and then finally withdrawn at midnight.

It was almost ten miles up the pike to Spring Hill where the column would join Stanley and the wagons; and the Confederate army was camped on the right side of the pike just before Spring Hill.

Cox started out at 7 P.M. Barring any obstacles, the head of the column should arrive in Spring Hill between 10 and 11 P.M. As the column approached Spring Hill, the troops were amazed to see hundreds, even thousands, of fires to the right of the pike coming to within a couple hundred feet of the pike. The column was not unnoticed. They passed Confederates singly and in groups staring at them, but the column continued on. According to General Cox, the march was unusually orderly. He stated:

> The march from Duck River to Franklin was made in the most perfect order; the men knowing that they were moving near the enemy's positions, kept well closed up.[4]

One of the observing Confederate soldiers, a private more enterprising than the others, rushed over to Hood's headquarters to report what he had seen.[5]

It was now about midnight and the fatigued, invalided Hood was once again awakened to hear the private out. Hood then wrote or caused to be written a note to General Cheatham, to be delivered by courier, to the effect that Cheatham should stop the movement on the pike. This was Hood's only written battle order of the night. Hood's and Cheatham's recollections of the contents and wording of the note were somewhat different. Let us first look at Hood's:

> About 12 P.M., ascertaining that the enemy was moving in great confusion, artillery, wagons and troops intermixed, I sent instructions to General Cheatham to advance a heavy line of skirmishers against him and still further impede and confuse his march.[6]

Cheatham's recollection of the note was:

> A courier from headquarters brought a note from Major A. P. Mason [Hood's A. A. G.] to the effect that General Hood had just learned that stragglers were passing along the road in front of my left, and "the Commanding General says you had better order your picket line to fire on them."[7]

Did the note thus actually identify the movement on the pike as enemy "stragglers?" During the Civil War, every large body of marching men covering any distance was always followed for hours afterward by stragglers. These were individuals who had to drop out of ranks for one reason or another and were now trying to catch up. If the note identified the Union movement on the road as stragglers (presumably from Stanley's column of the morning of the 29th), this was far different from identifying the traffic on the road as the main Union force from Columbia. Dealing with stragglers would be a minor matter. Furthermore, the note indicated that the traffic should be stopped with either a picket line or a line of skirmishers. This again implied that stopping the traffic on the road was a minor matter to be handled by a small number of troops.

Cheatham read the note and decided that the matter should be handled by the command bivouacked closest to the road. This was the division of General Johnson. Cheatham thus sent the note by Major Bostick of his staff to Johnson, along with the additional instructions that Johnson was to use any means necessary to stop the movement. Major Bostick later informed Cheatham that upon being awakened and receiving the order, General Johnson commenced complaining bitterly about being loaned out, and asked why General Cheatham did not use one of his own divisions, rather than that of Johnson, who was subordinate to Stewart and not Cheatham. At length, however, Johnson ordered his horse and he and Bostick rode down to the pike. It was after 2 A.M. before they reached the pike, over two hours since Hood issued his order, even though Hood, Cheatham, and Johnson were all within a mile of each other. Johnson and Bostick found the pike quiet and Johnson returned to bed.[8] Unknown to Johnson and Bostick, the rear guard of Schofield's column had already passed on its way to Spring Hill.

Schofield was active throughout the night. His infantry cleared Forrest from Thompson's Station, and the pike was now open all the way to Franklin. His column kept moving throughout the night and by 5 A.M. the head of the column had reached Franklin and the last of the 800 wagons was on the pike one mile beyond Spring Hill.

Had Hood's order to Cheatham at midnight simply stated, "The main Union force is on the pike moving toward Spring Hill. Stop it immediately," it might have all turned out differently. Cheatham's troops might have hit the moving Union column in the flank before it reached Spring Hill, and Hood would have had his victory. But it was not to be.

When the Confederates woke up on the morning of the 30th, the Union column was already disappearing in the distance down the pike toward Franklin.

Hood was furious when he learned that the Union troops had passed through his camp during the night and ordered an immediate pursuit.

When Schofield reached Franklin, he did not encounter the reinforcements he had been given to expect. However, he received something even more valuable — a prepared defense position. Franklin was in a bend of the Harpeth River, and a year earlier, defensive earthwork positions had been prepared for another battle that closed off the bend. There were two bridges at the center of the bend continuing on north toward Nashville: a regular bridge that had been destroyed, and a railroad bridge that was intact. Schofield, by placing planking over the railroad bridge, was able to cross his artillery, which was then placed on the far side of the Harpeth River to dominate his earthwork defense positions across the arc. Any frontal attack against the earthworks would thus be enfiladed by the artillery on the far side of the river.

Schofield's defensive position at Franklin was extremely strong and any frontal attack would be doomed to disastrous losses. By the time Hood arrived at Franklin early on the

30th, Schofield was ensconced in his earthworks and his artillery was positioned on the far side of the river.

Hood's generals took one look at the situation and were almost unanimous in their recommendation that Hood should not attack. After all, up to this point Hood had lost nothing but an opportunity, and his force was still intact. Hood, however, would have none of it. He could have, should have, and would have won a smashing victory in the last 24 hours had it not been for avoidable missteps of subordinates. Now, he was determined to retrieve an opportunity that was already forever lost. The same psychology that induced Lee to attack at Malvern Hill and to order Pickett's charge at Gettysburg overcame him. He would throw the dice in an attempt to recover a lost opportunity in a gamble he would never have taken at the outset. He would make an all-out frontal attack.

The outcome was predictable. The heart and soul of Hood's army was destroyed. Six of his best generals were killed, including the irreplaceable Patrick Cleburne, and five were wounded. After the battle, Schofield successfully withdrew to Nashville where he united with Thomas.

Any chance of victory for Hood was now gone. With his damaged, inferior force he foolishly laid siege to Nashville. His army after Franklin was a shell of itself and had in it only one more fight. This was to come soon.

On December 15, Thomas came out of Nashville and utterly destroyed Hood's besieging force. The war in the west was substantially lost for the Confederates. The seeds of Hood's destruction were sown at Spring Hill, sprouted at Franklin, and harvested at Nashville.

Had Hood written a better battlefield order the night of November 29–30, it might all have worked out differently. He should have written, "The whole Yankee army is marching down the pike to Spring Hill. Stop them immediately."

13

The Battle of Five Forks

The Situation Up to Mid-March 1865

Up to March 1864, the war in the east was give and take. The South would appear to be winning, then the North, then the South, and on and on. It appeared that there would be no end; that the war would go on ad infinitum. In desperation, Lincoln sent to the west for Grant, his thus far most successful general. In March 1864, he appointed Grant commanding general of all the armies. Grant, deciding that the decisive theater was the east, made his home with the Army of the Potomac and became its de facto commanding general, even though its titular commanding general, General Meade, continued in office.

Things were now different. Grant seized the initiative and was never to relinquish it. He pushed on south and, whatever the problems and whatever the casualties, continued on; there was now to be no turning back. Grant was resolved to seize Richmond, smash Lee, and end the war.

Despite all Grant's determination and efforts, he was not able to beat his way into Richmond. He was compelled to circle around it to the east, and by mid–June 1864, was facing the southern suburb of Richmond, Petersburg, from the south. We had now reached the ludicrous situation of Grant's northern army facing north, and Lee's southern army facing south.

Grant's new position to the south of Richmond was not all bad; in fact, it was good. Richmond's greatest vulnerability was from the south. All of the supplies needed to feed Richmond's inhabitants and Lee's army flowed in from the south and southwest. If Grant could cut these off, Petersburg, Richmond, and Lee's army would starve.

At this point, the nature of the war in the east changed, from mobile warfare to siege warfare. Lee's southern army dug in. It was well known by this time that one properly dug in man could defend against several in the open; and the Confederates became, indeed, well dug in. Their entrenchments became extensive and ingenious.

Grant's army at this time was twice as large as Lee's, but Grant, looking at the Confederate entrenchments, knew that he could not take them by storm. Just weeks earlier, on June 2, 1864, Grant had attempted to storm a Confederate entrenched position at Cold Harbor while the two armies were still contesting, to the north and east of Richmond. In this undertaking, he had suffered approximately 7,000 casualties in minutes while achieving no particular gain. The Confederate entrenchments he now confronted were even more for-

13. *The Battle of Five Forks* 163

Map 31
Five Forks — Situation on March 31

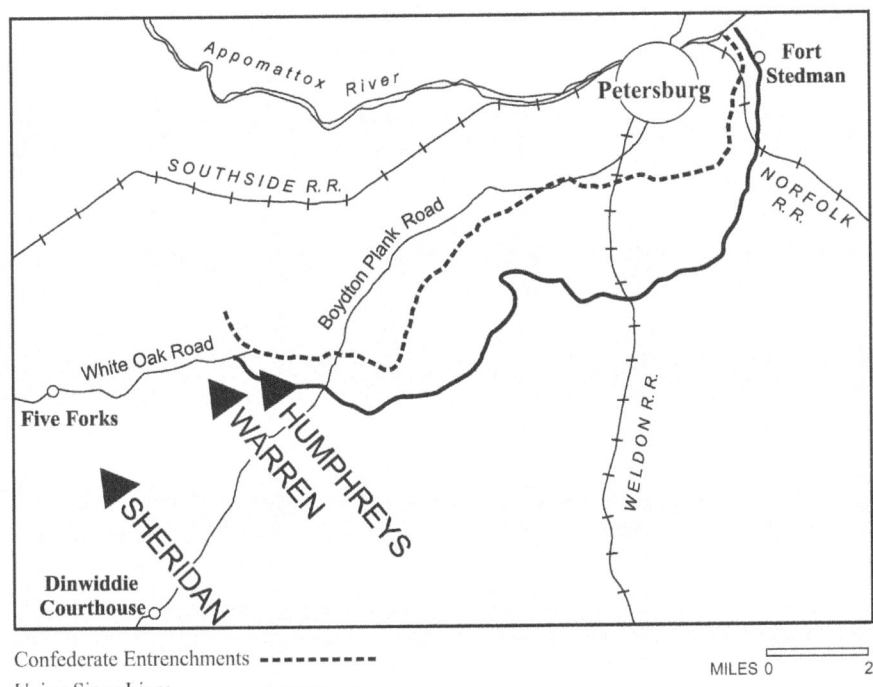

Map 32
Five Forks — Rail Lines into Richmond from South and West

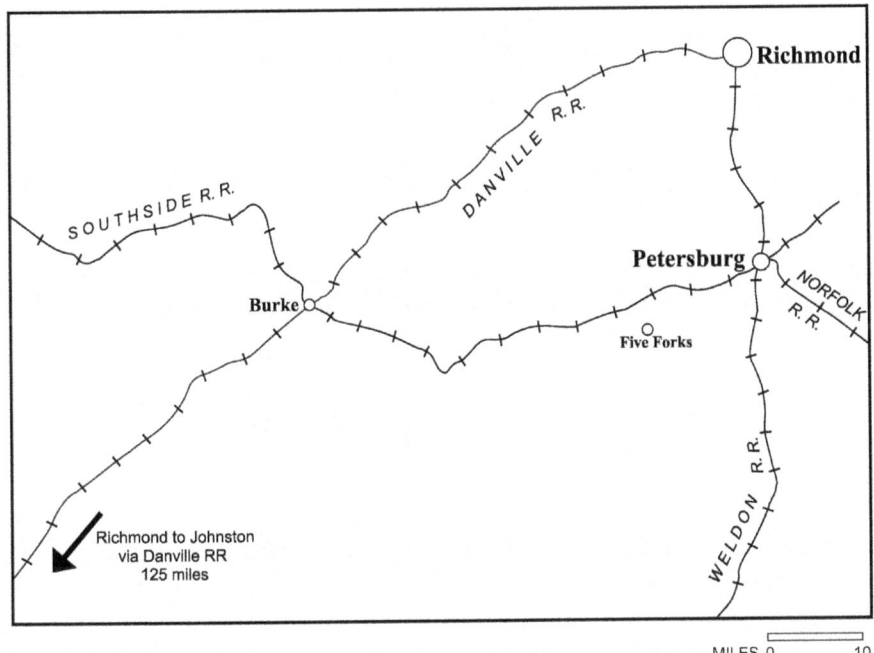

midable, and Grant knew that, by some means, he would have to thin out the defenders before attempting to move north and seize Richmond. He would not again make the mistake he had made at Cold Harbor.

Grant resolved, with his superior numbers, to push his siege lines ever further to the west. This would require the Confederates to extend their fortifications ever further west and thus thin out the number of men for manning each mile of fortification. Another advantage of Grant extending his siege lines ever further to the west was that he would be cutting the supply lines into Petersburg-Richmond from the south one by one.

By mid–March 1865, Grant's siege lines extended as far west as the Boydton Plank Road (see map 31). The Weldon Railroad into Richmond had been cut, and the only other rail lines into Richmond were now the Southside Railroad from the southwest and the Danville Railroad from the south (see map 32). The two rail lines crossed at a place called Burke, 45 miles west of Petersburg.

The siege had now been in progress for ten months, the Confederate entrenchments still appeared too strong to storm, and the public was tiring of a war that, it appeared, could go on forever.

While the siege of Petersburg-Richmond was in progress, the Confederates suffered unmitigated disasters in other theaters. General Sheridan, the Union commander of the Shenandoah Valley Army, had decisively defeated his Confederate opponents, cleared the valley for the Union once and for all, and was now en route to join Grant. Farther west, the Union army under Schofield grievously wounded the Confederate army under Hood at Franklin, and Union General Thomas completed its destruction at Nashville in December 1864. General Sherman's Union army had crossed Georgia from west to east, and was now pushing up the Atlantic coast through the Carolinas to join Grant. The only major Confederate fighting force still in the field between the Atlantic and the Mississippi besides Lee's army was a scratch army under General Joseph Johnston that was attempting to arrest Sherman's progress through the Carolinas (see map 32).

There was one last hope for Lee — slight, but just possible. If Lee could disengage his army from Grant, that is, slip away and steal a march, and be transported south by the remaining railroads and join Johnston, the now combined armies of Lee and Johnston just might be able to defeat Sherman. The combined Confederates could then turn to deal with Grant. Grant was very much aware that as long as the rail lines were open to Lee, Lee just might pull the rabbit out of the hat. Grant thus had two good reasons for cutting the last rail connections with Richmond: (1) to starve out Lee and (2) to prevent Lee from uniting with Johnston. We now come to the critical days of March 25 to April 2 that determined the outcome of the war.

The Critical Eight Days — March 25 to April 2, 1865

By mid–March, Lee knew that he had to act and had to act soon. If he sat tight behind his fortifications, Grant would ultimately cut the last rail lines out. He would no longer be able to unite with Johnston and he would starve. In order to disengage his army from Grant and withdraw, he would have to strike a blow sufficient to temporarily immobilize Grant. To this end, he secretly thinned out his entrenchments and accumulated a strike force. The point of attack would be Fort Stedman (see map 31) and the time of the attack would be the early morning hours of March 25. The attack seemed to succeed initially, but then faltered.

We might compare Lee's attack with that of the Germans in the Ardennes in the closing days of World War II that resulted in the "Battle of the Bulge." Both were desperation measures and, in both instances, the attacker simply lacked the means to carry it through; in both instances, it simply resulted in the quickening of the end. The end result of Lee's attack was 4,000 Confederate casualties versus 1,500 for the Union.

General Sheridan reported in to Grant on March 27 with his valley army. He brought two cavalry divisions and an infantry corps to the table. Grant believed that, with Lee's losses at Fort Stedman, he now had the means to bring the war to an end. The plan he formulated had three elements as follows: (1) He would extend his siege line further west to the White Oak Road, thus causing Lee to further attenuate the manning of his fortified line. He would accomplish this by pulling the corps of Humphreys and Warren out of line, moving them westward and replacing them with Ord's corps that was presently north of the James. (2) He would augment Sheridan's two cavalry divisions with a third, and then the divisions under Sheridan would move westward well south of the Union siege line and cut the last rail lines to Richmond. Grant anticipated that Lee would detect Sheridan's movement and would be required to remove men from his entrenchments into the open and march south to try to stop Sheridan. (3) Whether or not Sheridan succeeded in cutting the rail lines, once Lee removed additional men from the entrenchments to stop him, the entrenchments would be so lightly held that Grant could now pierce them with an all-out attack. This all-out attack should come on or about April 2–3.

Once the Confederate line of entrenchments was pierced and Grant entered Petersburg and then Richmond, the Confederates remaining in the entrenchments would have to be withdrawn and it would now be just a matter of mopping up. This should finally end the war. The crucial period would be the eight or nine days from March 25 to April 2 or 3.

Grant Issues His Orders

Before we look at the orders Grant issued to implement his plan, let us take a brief look at how the command situation had improved since the early days of the war. First of all, the use of the telegraph was greatly expanded. Throughout the unfolding operation, Grant's headquarters and Meade's headquarters were able to maintain telegraphic contact with Sheridan and the corps commanders. This, of course, should not be understood to mean that both Grant and Meade could always instantaneously reach Sheridan and the corps commanders personally. Each telegraphic terminal was set up at the army or corps headquarters of the moment, and the active commander could be miles away on horseback at the moment a message came in.

Second, staffs were now larger, more experienced, and better used. Grant always kept his staff informed of his plans, intentions, and desires, and visits between his staff members and the implementing commanders were frequent during an operation.

Lastly, Grant, always a good order writer, now had four years of experience.

Grant met personally with Sheridan on the 28th and explained his plans to him. He then handed Sheridan his orders in writing. The orders were:

City Point, VA., March 28, 1865
Maj. Gen. P. H. Sheridan,
Commanding Middle Military Division:

    The Fifth Army Corps will move by the Vaughan road at 3 A.M. to-morrow morning. The Second moves at about 9 A.M., having but about three miles to march to reach the point designated for it to take on the right of the Fifth Corps after the latter reaching Dinwiddie Court-House. Move your cavalry at as early an hour as you can and without being confined to any particular road or roads. You may go out by the nearest roads in rear of the Fifth Corps, pass by its left, and passing near to or through Dinwiddie, reach the right and rear of the enemy as soon as you can. It is not the intention to attack the enemy in his intrenched position, but to force him out if possible. Should he come out and attack us or get himself where he can be attacked, move in with your entire force in your own way and with the full reliance that the army will engage or follow the enemy as circumstances will dictate. I shall be on the field and will probably be able to communicate with you. Should I not do so, and you find that the enemy keeps within his main intrenched line, you may cut loose and push for the Danville road. If you find it practicable, I would like you to cross the South Side road between Petersburg and Burkeville and destroy it to some extent. I would not advise much detention, however, until you reach the Danville road, which I would like to strike as near to the Appomattox as possible. Make your destruction on that road as complete as possible. You can then pass on to the South Side road west of Burkeville and destroy that in like manner. After having accomplished the destruction of the two railroads, which are now the only avenues of supply to Lee's army, you may return to this army, selecting your road farther south, or you may go into North Carolina and join General Sherman. Should you select the latter course, get the information to me as early as possible, so that I may send orders to meet you at Goldsborough.
    U.S. GRANT,
    Lieutenant-General[1]

    This was not one of Grant's clearest orders. It advised Sheridan of the movements of the Fifth Army Corps (Warren) and the Second (Humphreys) to his north, but never required him to contact or be in communication with either. It did tell him not to attack the Confederates if they remained in their entrenchments, but to attack them if they came out. Also, if he did that, he would be supported.

    As to where he was to go, it only told him to "reach the right rear of the enemy as soon as you can" and in doing so to pass through or near Dinwiddie Courthouse. He was further directed that, if the enemy "keeps within his main intrenched lines," he may "cut loose and push for the Danville Railroad" and "if practicable," destroy both rail lines. Thus we may summarize: (1) He was to reach the right rear of the enemy as soon as possible. (2) Do so by passing through or near Dinwiddie Courthouse. (3) Attack the enemy if he came out of his entrenchments, but not to do so if he did not. (4) If the enemy did not come out of his entrenchments, he was to "cut loose" and destroy the two railroads leading to Petersburg-Richmond.

    A careful reading indicates that Sheridan was tasked with destroying the railroads only if Lee did not come out his entrenchments to confront him. The main objective was to use the threat to destroy the railroads as a means to draw Lee out of his entrenchments.

    If Sheridan wanted to reach the Southside Railroad, he had to continue northwest on the Dinwiddie road for four miles past the courthouse. At that point he would reach an

13. The Battle of Five Forks

insignificant settlement called Five Forks. Five Forks was naturally the junction of five roads. The fork to the northeast was called White Oak Road. Just four miles up White Oak, one came to the western terminus of the Confederate defense line. If one continued directly through Five Forks to the northwest, the name of the road changed from Dinwiddie Road to Ford Road. Just two miles up Ford Road, one came to Ford's Crossing on the Southside Railroad. Thus, once Sheridan reached Dinwiddie Courthouse, he had just six miles to go to the railroad. He would have to follow the roads to get to the railroad, as there was no possibility of deploying across the soggy fields (see map 33).

If the Confederates wanted to stop Sheridan from reaching the railroad, they would have to march down White Oak Road to Five Forks. From here they would have to establish

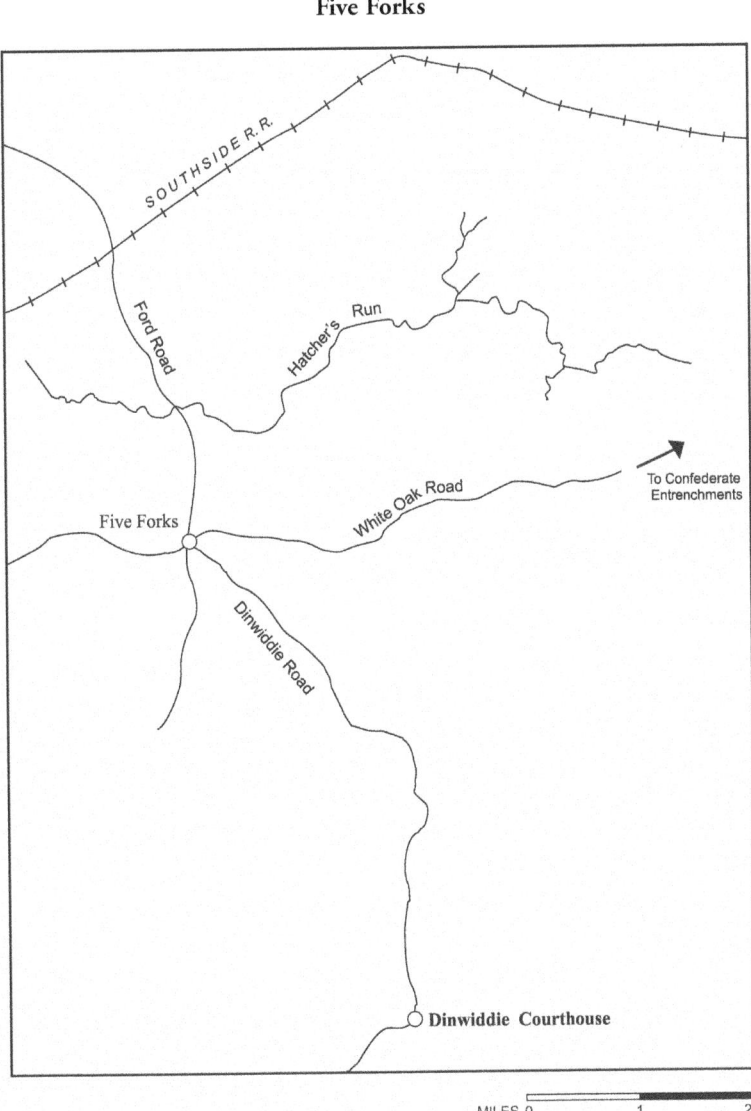

Map 33
Five Forks

a blocking position either at Five Forks itself or somewhere on the two-mile stretch of Fords Road between Five Forks and the railroad.

Establishing the blocking position at Five Forks itself would have the advantage of keeping open the possibility of maintaining contact with their main defense line four miles up White Oak Road. However, the disadvantage would be that there were no natural obstacles at Five Forks on which to anchor the line.

An alternative possibility for establishing a defense line was to establish it behind Hatcher's Run. Hatcher's Run crossed Ford Road at right angles approximately midway between Five Forks and Ford's Crossing. It was now in full flood and, properly manned, could provide a formidable defense position. The disadvantage would be that, if the Confederates made their stand here, they would be isolated from their main defense line.

Lee learned of Sheridan's movements March 29, soon after Sheridan started and well before Sheridan reached Dinwiddie Courthouse. Lee naturally assumed that Sheridan's objective must be the railroads, and concluded that he must be stopped at all costs.

After Lee's losses at Fort Stedman, he had fewer than 40,000 men to man almost 30 miles of entrenchments, and Grant had almost four times that number. Lee's options were thus severely limited. He scraped up what forces he could. These consisted of three cavalry divisions under General Fitzhugh Lee consisting of about 5,000 men, and Pickett's infantry division of three brigades augmented by two brigades of Johnson's division, for a total of another 5,000 men. The total force was thus almost 10,000, and the overall commander was Pickett. The cavalry reached Five Forks first, and the infantry set out to join them on the 30th.

In the meantime, the Union corps of Humphreys and Warren were moving west to extend the Union siege line, and Warren actually arrived close enough to the White Oak Road on the 30th to see Pickett's division marching down the White Oak Road toward Five Forks.

The ball was about to open. It would open March 31, 1865. The weather and terrain were to have a profound effect on what followed, so we will digress here a bit to look at these aspects.

The Weather and Terrain

In the words of Horace Porter, a member of Grant's staff, they were "in a section of country in which the dust in summer was generally so thick that the army could not see where to move, and the mud in winter was so deep that it could not move anywhere."[2]

As of March 29, when Sheridan set out to implement his orders, they were more on the mud side than the dust side. Grant realized that the drying of the roads was just as much an advantage to Lee as to him; in fact more so, as he feared that Lee would flee as soon as the roads dried. Consequently, he elected to begin his campaign in the chancy intermediate period.

General Warren described the terrain he was to pass through as follows:

> The country in which we were to operate was of the forest kind, common to Virginia, being well watered by swampy streams. The surface is level, and the soil clayey or sandy, and where these mix together, like quicksand. The soil, after the frosts of winter first leave, is very light and soft, and hoofs and wheels find but little support.[3]

As luck would have it, torrential rains fell in the campaign area the night of the 29th and all day on the 30th. By the evening of the 30th, whole fields had become beds of quicksand in which horses sank to their bellies, wagons threatened to disappear altogether, and it seemed that the bottom had fallen out of the roads. The men began to feel that if anyone in later years asked them if they had had been through Virginia, they could answer, "Yes, in a number of places."

The roads in the area were few and unpaved and it was obvious that operations would be restricted to the roads, such as they were. The terrain was interlaced with swampy streams that were now overflowing. Some could be crossed by infantry with difficulty, but essentially all required bridging to be crossed by wagons or artillery.

March 31, 1865 — The Opening Moves

March 31, 1865 was the beginning of the end. The Union corps of Warren and Humphreys were pushing westward just below the Petersburg entrenchments to extend the Union siege line and thus force the Confederates to extend the defense lines and thin out the whole. As of March 31, Warren-Humphreys had reached the White Oak Road and were less than two miles from the current western terminus of the Confederate line.

A little over five miles to the south of the troops of Warren and Humphreys was the cavalry force of General Sheridan. Sheridan was also moving westward toward the vital Southside Railroad, a critical supply line for the Confederates. As of the 31st, he was already pushing beyond Dinwiddie Courthouse and was less than five miles from the railroad.

As of the 30th, Lee had dispatched a mixed cavalry and infantry force of 10,000 under General Pickett down the White Oak Road to Five Forks, the last road junction between Sheridan and the railroad. We thus had two separate threats against the Confederates: the Warren-Humphreys force to the north, and the Sheridan force to the south. They both had to be stopped and they both had to be stopped now, today, March 31.

The two Union threats were separate and distinct. Although the two forces were within cannon sound of each other and moving in the same direction, they were separated by Gravelly Run, which at the time could be crossed only by bridge, and the bridge in the Union section was then down. Furthermore, the two forces were under the overall command of two separate commanders and their staffs. The Warren-Humphreys force to the north reported to General Meade, the commanding general of the Army of the Potomac; and the Sheridan force to the south reported directly to General Grant, the overall Union commanding general.

Telegraphic communications required a wire between the correspondents, but were far advanced on the Union side by this stage of the war. Union corps had road mobile telegraphic stations, and wire was strung as the corps advanced. The telegraphic terminals available to the Union forces on March 31, 1865, are as indicated in figure 2. Warren and Humphreys were connected to Meade's headquarters, Meade's headquarters to Grant's headquarters, and Grant's headquarters to Sheridan's. This telegraphic network gave the Union forces a huge advantage. There was a constant flow of orders and reports via the network during the day and night of the 31st, allowing the Union forces to adapt to a changing situation.

Pickett's force at Five Forks, on the other hand, had no telegraphic connection with

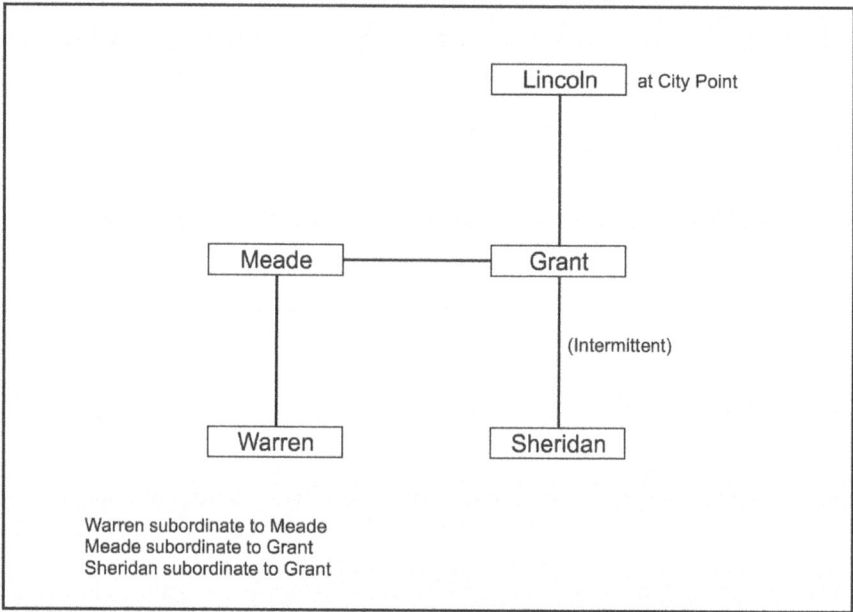

Figure 2
Five Forks — Telegraphic Connections of the Union

Lee and, once Warren arrived at the White Oak Road, even his primary road connection with the Confederate main lines had, at least temporarily, been cut. He could only receive orders or send reports via courier, and then by a circuitous route. Pickett and his force of 10,000 were isolated; cut off from the main Confederate lines.

The Confederates went on the attack in both theaters on the 31st: in the north to stop Warren-Humphreys, in the south to stop Sheridan.

In the south, Pickett's force of 10,000 was about equal to that of Sheridan, and he managed to push Sheridan back to Dinwiddie Courthouse. In the north, the fighting was back and forth. There was no coordination between the Union forces in the north and south; it was two separate and distinct battles.

In the evening, Sheridan telegraphed the following message to Grant:

Headquarters Cavalry
Dinwiddie Court House, March 31, 1865
Lieutenant-General Grant, Comdg, Armies of the United States

    The enemy's cavalry attacked me about 10 o'clock to-day on the road comming in from the west and a little north of Dinwiddie Court-House. The attack was very handsomely repulsed by General Smith's brigade, of Crook's division, and the enemy was driven across Chamberlain's Creek. Shortly afterward the infantry attacked on the same creek in heavy force, and drove in General Davies' brigade, and advancing rapidly, gained the forks of the road at J. Boisseau's. This forced Devin who was in advance, and Davies to cross to the Boydton Road. General Gregg's brigade and General Gibbs' brigade, who had been toward Dinwiddie then attacked the enemy in the rear very handsomely. This stopped the march toward the left of our infantry, and finally caused them to turn toward Dinwiddie and attack us in heavy force. The enemy then again attacked at Chamberlain's Creek and forced Smith's position. At this time, Capehart's and Pennington's brigades of Custer's division, came up and a very handsome fight occurred. The enemy have gained some ground, but we still hold on in front of

Dinwiddie, and Davies and Devin are comming down the Boydton Road to join us. The opposing force was Pickett's division, Wise's independent brigade of infantry, and Figzhugh Lee's, Rosser's and W. H. F. Lee's cavalry commands. The men have behaved splendidly. Our loss in killed and wounded will probably number 450 men. Very few were lost as prisoners. We have of the enemy a number of prisoners. This force is too strong for us. I will hold out to Dinwiddie Court House until I am compelled to leave. Our fighting today was all dismounted.
P. H. Sheridan
Major General[4]

Sheridan's identification of the Confederate units confronting him was right on. However, the key sentence of his whole message was, "This force is too strong for us."

Grant responded at 10:05 P.M. with the following:

Dabney's Mills, Mar 31, 1865–10:05 P.M.
Major General Sheridan:
The Fifth Corps [Warren] has been ordered to your support. Two divisions will go by J. Boisseau's and one down to Boydton road. In addition to this I have sent MacKenzie's cavalry, which will reach you by the Vaughan Road. All these forces should reach you by 12 tonight. You will assume command of the whole force sent to operate with you, and use it to the best of your ability to destroy the force which your command has fought so gallantly today.
U.S. Grant
Lieutenant General[5]

This was one of Grant's finest decisions, if not the finest of the war. He now gave Sheridan but a single task—that of destroying the enemy in front of him. He also gave Sheridan a blank check as to how to do it, and provided him the means to do it.

If Sheridan could eliminate Pickett's 10,000, these, plus the 5,000 Lee lost on the 25th, would deprive Lee of the means of manning 30 miles of fortifications plus protecting the railroads. The 10,000-man Pickett force that Lee had sent down the White Oak Road was sufficient to confront Sheridan. It was not nearly sufficient to cope with Sheridan and Warren combined.

There were some things that Grant did not know. Gravelly Run, between Warren and Sheridan, could not be crossed except by bridge, and the bridge was down. The order went out to Warren to march immediately to join Sheridan at Dinwiddie Courthouse with his three divisions. Warren spent the night furiously working on the bridge, but the first of his three divisions was not to reach Sheridan before daylight on April 1.

Warren also had to disengage from the enemy. As Warren withdrew, the White Oak Road again was opened to the Confederates and, at least for some hours, could have been used to reinforce Pickett via Five Forks. Warren anticipated that Confederate cavalry would follow him to see where he was going. He reasoned that, once the Confederates realized that he was going to reinforce Sheridan, they would use the now opened White Oak Road to reinforce Pickett. However, to Warren's amazement, no one followed him, and the Confederates made no effort to use the road while it was open. By morning it was too late. MacKenzie's cavalry, now subordinate to Sheridan, occupied the road. Warren was later to write:

It was a matter of wonder at the time, and has been ever since, how the enemy permitted our thus withdrawing without following us up to see the way we took, even if it had been with only a regiment. He would thus early have gained the knowledge that our infantry was moving toward the detached force under General Pickett, which we beat so badly toward evening. General Lee could then have re-enforced his detached troops or timely warned them to withdraw.... It was a want of vigilance that was most rare on their part and betokened that apathy which results from a hopelessness as to the use of further resistance.[6]

Grant was to delegate one more authority to Sheridan the night of March 31–April 1. He authorized Sheridan to remove Warren from command and replace him with one of Warren's division commanders if he so desired. This action was un–Grant-like. Why did he do it? Warren was certainly not stupid nor a coward nor incompetent. In fact, he had indisputably made contributions to winning the war up to this point. Warren was, however, no firebrand. He was cautious and methodical.

Grant could now smell the end of the war. The possibility of achieving his most cherished dream of bringing the war to an end was at hand. He must seize the moment. The moment required action and decisiveness. Sheridan was a can-do man, a hell-for-leather firebrand. Warren was not. Sheridan must not be restricted in any way at this decisive junction. If need be, Warren would have to go.

Pickett learned of Union troops above moving south during the night. He feared that when the sun came up he would find Union troops behind him, between him and the railroad. He chose to move back toward Five Forks.

Events were now moving inexorably toward the time and place of the final showdown — Five Forks on April 1, 1865.

Before we go to April 1, let us take a brief look at the key players in this drama.

The Leading Players

The leading players in this drama were George Pickett on the Confederate side and Phil Sheridan and Gouverneur Warren on the Union side. We will start with Pickett.

Strangely enough, George Pickett was appointed to West Point in 1842 by none other than congressman Abraham Lincoln. Young Pickett graduated from West Point in the famous class of 1846. That is, just barely; he finished last in a class of 56. This justly caused many to conclude that he was not very bright. The class of 1846 provided more generals to the Civil War than any other class. These included such luminaries as George McClellan, John Gibbon, Jesse Reno, George Stoneman, and Samuel Sturgis on the Union side, and A. P. Hill, Stonewall Jackson, and Cadmus Wilcox on the Confederate side.

The class graduated in time to play a prominent part in the Mexican War, in which Pickett served with distinction. He received two brevets and achieved prominence for triumphantly unfurling the U.S. flag over the ramparts at the battle of Chapultepec.

At the outset of the Civil War, he had 15 years of experience as an officer in the regular army and, thus, quickly obtained a commission as colonel in the Confederate army. He was promoted to brigadier general in February 1861, and major general in the fall of 1862. Pickett's performance as general can best be described as pedestrian.

His real claim to fame and immortality occurred at the battle of Gettysburg. Lee ordered a foolish charge of 15,000 men across one mile of open ground against Union troops ensconced behind a stone wall. This was forever after referred to as "Pickett's Charge."

13. The Battle of Five Forks

When the average American visits the Gettysburg battlefield and hears of Pickett's Charge for the first time, he is apt to visualize ranks of charging men with a mounted Pickett out in front gallantly waving his sword. This, of course, is all nonsense. The charge involved three divisions, of which Pickett commanded only one. The charge was largely covered and reported upon by reporters from Virginia newspapers, and Pickett's division alone consisted of all Virginians. Thus, these got the lion's share of the publicity. The overall casualties suffered by the charging 15,000 were 55.4 percent. Two of the three division commanders were casualties, as were all three of Pickett's brigade commanders, and 11 of his 15 regimental commanders. However, neither Pickett nor any of the staff officers who were with him suffered a scratch. Exactly where were they during the charge? No one is sure to this day. Perhaps hiding behind some rock? Pickett, on the other hand, spoke very highly of his participation in the charge in a letter he wrote to his fiancée shortly after the event. Pickett wrote:

> Even now I can hear them cheering as I gave the order, "Forward!" I can feel their faith and trust in me and their love for our cause. I can feel the thrill of their joyous voices as they called out all along the line "We'll follow you, Marse George. We'll follow you, we'll follow you." Oh, how faithfully they kept their word, following me on, on to their death, and I, believing in the promised support, led them on, on, on. Oh God![7]

Be that as it may, when the charge is mentioned, or Pickett's name is mentioned, one tends to think of the gallant Pickett. Pickett next appeared front and center in the Civil War at Five Forks.

We will next turn to Union Major General Gouverneur Warren, commanding general of the Fifth Corps. If General Pickett's acumen was suspect, General Warren's was not. Warren entered West Point at the age of 16 and graduated in 1850, second in a class of 44. He was then commissioned into the prestigious Topographical Engineers.

Almost all of Warren's pre-war duties consisted of nonmilitary engineering duties. He participated in studies for transcontinental railroad routes, conducted surveying of the routes, and significantly contributed to the mapping of the west.

He was appointed as lieutenant colonel in the volunteer army at the outset of the war, quickly advanced to brigadier general, and proved a competent and reliable combat commander. However, in February 1863, because of his engineering background, he was appointed chief engineer of the Army of the Potomac. It was in this capacity that he is today best remembered.

On July 2, 1863, during the battle of Gettysburg, Warren, himself not commanding any troops, noticed that Little Round Top was the key to controlling the battlefield and was unoccupied. He took it upon himself to order the nearest Union command to divert from its orders and occupy the crest. This was accomplished only moments before a Confederate unit arrived to occupy the hill. Warren thus saved the day. Had he not so acted, there is a good chance that the Confederates would have won the battle. Today, there is a giant statue of Warren atop Little Round Top, gazing out over the battlefield.

Warren was promoted to major general in August 1863, and resumed serving as a combat unit commander. He was initially placed in command of the Second Corps, but then transferred to command the Fifth Corps. Warren again proved himself a competent and reliable, albeit cautious, combat unit commander. It was in his capacity as Fifth Corps commander that we pick up his story at Five Forks.

When one reads Warren's post operations reports today, one cannot but conclude that he was an unusually intelligent, clear-thinking, and thorough man.

Of our three leading players, Sheridan was by far the best known to the public in 1865. In the North, he was a swashbuckling hero; in the South, a dastardly villain second only to Sherman. Sheridan, like the other two players, was a West Point graduate, graduating 34 out of 52 in the class of 1853.

The outset of the war in 1861 found him a lowly first lieutenant with no particular prospects or recognition. For his first wartime assignment, he was assigned to work directly for Major General Henry Halleck, then head of the Department of Missouri, headquartered in St. Louis. Halleck put Sheridan to work auditing the accounts of his predecessor. Halleck was greatly impressed with Sheridan and his work (as a bookkeeper) and to an extent became his patron.

Sheridan finally secured an opportunity to obtain a combat unit command in May 1862, when he was appointed colonel of the 2nd Michigan Cavalry. Here, Sheridan discovered his true forte, that of a charismatic battlefield commander. His superiors were so impressed that they quickly forwarded a recommendation to Washington to promote Sheridan to brigadier general. The recommendation passed over the desk of none other than General Halleck, who was now general-in-chief of all the armies. Naturally, it was approved, and Sheridan was on his way. Sheridan went from success to success in the west and was promoted to major general in April 1863. Sheridan came to Grant's attention and, in the spring of 1864 when Grant was called east to head all the armies, he sent for Sheridan.

Confederate operations in the Shenandoah Valley had been a thorn in the side of the Union since the beginning of the war. Sheridan did what no other Union general had been able to do. He ended the Confederate Shenandoah operations once and for all. He destroyed the Confederate Shenandoah army, burned the crops, carried off the livestock, and destroyed the infrastructure.

Sheridan's success brought adulation in the press and the famous poem, "Sheridan's Ride," was written and became immensely popular. It described Sheridan rallying his fleeing troops after they had been temporarily defeated at the battle of Cedar Creek. In a probably apocryphal addition, we have the mounted Sheridan shouting to Union troops running past, away from the action, "Remember men that you are from Ohio." One man rushing past responded, "Yes, by God, and I am trying to get back there as fast as I can."

We will terminate our brief article on Sheridan with a description of him provided by Abraham Lincoln, "A brown, chunky little chap, with a long body, short legs, not enough neck to hang him, and such long arms that if his ankles itch, he can scratch them without stooping."[8]

We will now return to the fateful day of April 1, 1865.

April 1, 1865 — The Battle of Five Forks

Before we turn to the Battle of Five Forks, let us digress a bit to give a short tutorial on "flanking." In the Civil War, men lined up shoulder to shoulder in a "line of battle" to fire at the enemy. The two ends of the line were called the flanks. Thus, every line of battle had a left flank and a right flank. If the two opposing lines faced each other and were of

Figure 3
Five Forks — Flanks and Flanking

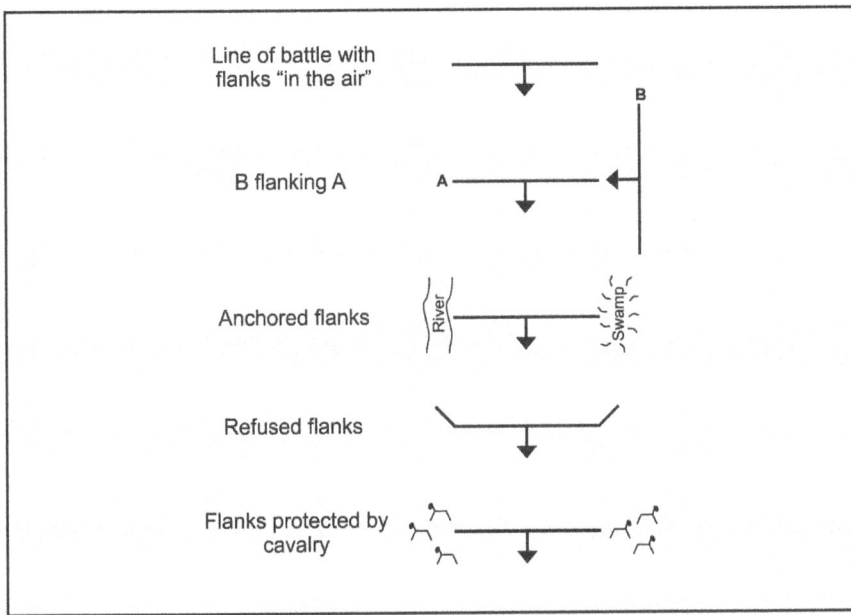

equal size, neither side had an advantage. However, if one side disappeared in the woods and then reappeared in a line of battle at a right angle to its opponent, it had "flanked" the opponent and had a decisive advantage (see figure 3). All of its guns would now be in the direction of the end man of the enemy line. This man would immediately realize that his situation was hopeless and run. Then, as the attacking line on the flank stepped forward, the next man of the enemy would be in the same hopeless situation and run, and so on and so on. Thus, the flanking army could literally "roll up" the line of its opponent.

In order to avoid being flanked, an army typically tried to "anchor" its flanks on some natural obstacle that the enemy could not cross. This could be a river, a swamp, etc. If the flank was not anchored, it was referred to as being "in the air."

If there were no physical features available to anchor its flanks, armies did the next best thing. They "refused" the flanks. This simply means that they bent back the extremities of their line of battle so that, if an enemy approached on the flank, at least some guns were facing it. Hopefully, this would provide time for the whole line to be reoriented to face the enemy. The angle of the refuse and its length were a matter of judgment and depended on the particulars of the situation at the moment. Cavalry was also commonly used to protect an otherwise open flank.

Five Forks was not a good place for the Confederates to fight a battle. There were no natural features on which Pickett could anchor his flanks while protecting Ford Road. Vastly superior forces were approaching him. He now had less than 10,000 men and Sheridan had an equal amount of cavalry plus over 12,000 of Warren's infantry. Of the five forks, only Ford Road radiated out behind him. The advantage of holding on to the White Oak Road as a connection to the main Confederate line was now gone. Union cavalry under General MacKenzie now occupied the road. Union forces could now approach Pickett at Five Forks by any or all of four of the forks. The area consisted of woods and fields and Union forces could move about hidden by the woods. It was all but certain that if Pickett made his stand at Five Forks, he could be flanked and overwhelmed.

There was a far better place for Pickett to make a stand between Sheridan and the railroad. That was behind Hatcher's Run. Hatcher's Run ran across Ford Road at right angles, about halfway between Five Forks and the rail crossing. Here, the attacking Union forces would have to cross a difficult obstacle under fire before even reaching the Confederate line. Here, there would not be four roads of approach to Pickett's position, but only one.

As Pickett retired toward Five Forks, he received an order from Lee via courier. It was, "Hold Five Forks at all hazards. Protect road to Ford's Depot and prevent Union forces from striking the Southside Railroad."[9]

Unlike Sheridan's orders, which gave him total freedom of action, Pickett's gave him none. The battle would be at Five Forks. Pickett sent his wagons down Ford Road beyond Hatcher's Run and dug in at Five Forks. His line ran for one and three-fourths miles along the north-south Forks, with Ford's Road protected and radiating out of the center of the rear of his line to the west. Neither flank of the line rested on any natural obstruction. He placed the bulk of his cavalry to protect his right flank (the one to the south) and refused his left flank (the one to the north). The refuse was at an angle of 90 degrees and extended westward 300 feet.

While work on the defense line was in progress, Confederate General Rosser, who had somehow acquired a supply of shad (tasty fish), invited Generals Pickett and Fitzhugh Lee, Pickett's cavalry commander, to a shad bake, which was to take place two miles up Ford Road behind Pickett's defenses. Pickett and Fitz Lee accepted and departed without notifying the next in command where they were going or when they would be back. Meanwhile, Sheridan was preparing his attack.

Sheridan's basic plan was as follows: His cavalry on foot would confront the length of the Confederate line. In the meantime, Warren's infantry corps would line up along White Oak Road out of sight of the Confederates. The divisions of Ayers and Crawford were to be side by side, Ayers on the left and Crawford on Ayers's right. Griffin's division was to line up behind. At the signal, the force was to pivot like a barn door with Ayers the left hinge. This would then place Warren's corps at right angles to Pickett's left flank. Then, while Sheridan occupied the Confederate line in front, Warren's force would provide the clincher and roll up the Confederate line.

It all sounds simple when heard and seen on paper, but was not so when implemented on the scene. The area was dotted with numerous patches of woods, and no one person could see the whole panorama of the battlefield. Each saw a different snippet depending upon where he was located. Then, too, the lines of battle were long. Warren's corps front of Ayers's and Crawford's divisions was three-quarters of a mile long. No one could even see the length of his own line.

Warren's force was slow in forming and the day moved on. His men had been fighting all day on the 1st after marching all night on the 31st-1st. None had had a dry place to lay down since the campaign started on the 29th, and the marching consisted of slogging in the mud. Most or all were dead tired.

Sheridan grew increasingly irritated as the day wore on. His cavalrymen had been confronting the Confederate front much of the afternoon, and he feared that they could run out of ammunition before the main ball opened. Furthermore, if he did not dispose of the Confederates before dark, he could not camp where he was and wait for the morrow. The ground he stood upon was soggy, and he would have to retire all the way to Dinwiddie Courthouse for bivouac.

To Sheridan's great annoyance, Warren was not ready to push off until after 4 P.M. through no fault of his own. Ayers's division hit smack into the Confederate refused sector, but Crawford and Griffin marched right past the end without seeing it.

Warren, quite properly for a corps commander, took up a fixed position in the rear. In this way, he could be found by messengers, receive reports, issue orders, and coordinate the various pieces that no one person could see.

Crawford and Griffin were gradually reoriented to hit the Confederate rear. Pickett and Fitz Lee returned to their leaderless command too late. The final result was absolutely predictable. The greatly outnumbered Confederates could not withstand being hit in front, rear, and left flank, and their position collapsed. Pickett's force was effectively destroyed, with the command being killed, wounded, captured, or scattered.

At 7 P.M., after the battle was already winding down, Warren received the following message from Sheridan:

Cavalry Headquarters
April 1, 1865
Major General Warren, Commanding Fifth Army Corps, is relieved from duty, and will report at once for orders to Lieutenant General Grant, Commanding Armies of the United States.
Jas. W. Forsyth
Brevet Brigadier General
and Chief of Staff[10]

By the time the sun set on April 1, 1865, the end of the war was in sight. Desertions in Lee's army increased. Nothing stood between Sheridan's cavalry and Lee's railroad lines and escape routes. Lee was down to fewer than 35,000 men, grossly inadequate to man 30 miles of entrenchments. Grant had over three times as many men. Grant attacked on April 2, and on April 3 Lee's lines were pierced. Lee's death throes lasted just six more days. The surrender came at Appomattox on April 9, 1865.

Pickett was never to receive another command. He was discharged from the army before Appomattox. He spent his remaining days as an insurance salesman and came to a premature death in 1875 at the age of 50. Pickett's claim to fame, or infamy as the case may be, was that he was the central figure in two of the South's greatest fiascoes, Pickett's Charge and Five Forks.

Pickett's young wife lived on for five decades after Pickett's death. She took to writing and lecturing, and the subject was her husband, whom she idolized. Largely because of her, when the name of George Pickett is mentioned today, it usually elicits visions of gallantry and greatness.

General Warren was not to learn of the reason or reasons Sheridan relieved him from command until he read it in the newspaper. There really was no one reason; it was a potpourri.

From a reading of Sheridan's official report of the battle, it would appear that he considered that Warren was remiss in all his actions from the very time he received orders to join Sheridan onward. He first criticized Warren for not getting in Pickett's rear during the night of the 31st while Pickett was before Dinwiddie Courthouse. He next criticized Warren for his slowness in bringing his corps into action on the 1st, and finally criticized Warren for his action in the battle itself. Let us here quote from Sheridan's report:

> In this connection I will say that General Warren did not exert himself to get up his corps as rapidly as he might have done, and his manner gave me the impression that he wished the sun to go down before dispositions for the attack could be completed.... During the attack I again became dissatisfied with General Warren. During the engagement portions of his line gave way when not exposed to a heavy fire, and simply from want of confidence on the part of the troops, which General Warren did not exert himself to inspire.[11]

Some of this sounds so nebulous that one might conclude that Sheridan had a prior grudge against Warren.

Why did Grant put Warren's fate in Sheridan's hands in the first place? Why did he explicitly, in writing, take the unusual step of authorizing Sheridan to relieve Warren? Let us turn to Grant's own words from his memoirs:

I was so much dissatisfied with Warren's dilatory movements in the battle of White Oak Road and in his failure to reach Sheridan in time, that I was very much afraid that at the last moment he would fail Sheridan. He was a man of fine intelligence, great earnestness, quick perception, and could make his dispositions as quickly as any officer, under difficulties where he was forced to act. But I had before discovered a defect which was beyond his control, that was very prejudicial to his usefulness in emergencies like the one just before us. He could see every danger at a glance before he had encountered it. He would not only make preparations to meet the danger which might occur, but he would inform his commanding officer what others should do while he was executing his move.[12]

When Warren appeared for reassignment, he was not treated as a pariah but, rather, given an important assignment commensurate with his rank. After the war, when the wartime giant U.S. Volunteer Army was disbanded, Warren reverted to his rank of major in the Topographical Engineers in the regular peacetime army. He served the next 17 years in various railway, harbor, and bridge study and survey projects. He died on August 8, 1882, at the age of 52. At the time of his death, he was still on active duty in the rank of lieutenant colonel.

Warren considered that he had been done an injustice and sought a court-of-inquiry. This request was repeatedly denied but finally granted in 1879. The court concluded that Sheridan's removal of Warren was unjustified. However, the findings were not published until after Warren's death.

General Sheridan had begun the war as a first lieutenant in the regular army and at its end was the fourth senior general in the regular army. This meteoric rise might be likened to an assistant bookkeeper in the Peoria Chevy dealership rising to CEO of General Motors in four years.

Sheridan was nationally famous at the end of the war and remained very much in the public eye afterwards. In 1867, Grant appointed him head of the Department of the Missouri with the task of pacifying the Plains Indians. The great Lakota War occurred during Sheridan's tenure. It was during this war that the greatest and most famous of all Indian battles occurred—Custer's Last Stand. Inasmuch as this was the last big battle under Sheridan's command, and that it involved all Civil War luminaries, we will cover it in this book. However, in that it did not occur during the Civil War, we will treat it as an appendix.

Sheridan went on to become commanding general, U.S. Army. He was appointed to this post on November 1, 1883, and served until shortly before his death on August 5, 1888, at the age of 57.

In addition to Sheridan's many military accomplishments, he is also credited with being largely responsible for establishing Yellowstone National Park.

The quote, "The only good Indian is a dead Indian," is also attributed to Sheridan. However, according to *Bartlett's Quotations*, that is not exactly what he said. According to Bartlett's, Sheridan's words were, "The only good Indians I ever saw were dead." According to Bartlett's, this remark was made at Fort Cobb, Indian Territory, in January 1869.

Sheridan has become an historic figure. His face has appeared on both U.S. currency and U.S. postage stamps. One city and five counties are named after him, as are schools, roads, a mountain, and a glacier.

Appendix: Custer's Last Stand

Custer's Last Stand may be considered the last hurrah of this work. Although it occurred 11 years after the Civil War, every one of the key players on the government side was a Civil War luminary and the conditions that we described still prevailed in spades. That is, commanders in a cooperating endeavor were widely separated, were not in communication with each other, were constrained in their actions by written orders that they carried, and had absolutely no opportunity to obtain clarification or elucidation. Inasmuch as this battle did not actually occur during the Civil War, but we find it relevant and interesting, we have relegated it to this appendix.

Our case is variously called "Custer's Last Stand" or the "Battle of the Little Big Horn." This battle occurred within the so-called Lakota War of 1876. A number of Indians of the various Sioux and Cheyenne tribes had refused to enter the reservations. In 1875, the Interior Department finally turned the matter over to the War Department for resolution. It was clear to all at this point that the matter could now only be resolved by fighting.

The noncompliant Indians were referred to as the "Hostiles." They roamed freely over an area of 90,000 square miles that encompassed much of Montana, Wyoming, and the Dakotas. Within this area there were no farms, ranches, or settlements of whites. The number of hostiles at the time the matter was turned over to the War Department was estimated as no more than 2,000 to 4,000, and the number of warriors no more than 500 to 800.[1] As the war began, the various hostile tribes were loosely united under the charismatic Indian medicine man Sitting Bull.

How to Find the Hostiles

The War Department's first problem, obviously, was to find the hostiles. They roamed freely within a 90,000 square mile area. Ninety thousand square miles translates into 57,600,000 acres. Even if all the tribes were united in a single village, the village might occupy less than 100 acres. Furthermore, the hostiles were hunters and gatherers who moved their village periodically as the local resources became exhausted. How then does one find a 100-acre, moving village in an expanse of over 57,000,000 acres?

There were two vulnerabilities. The first was water. The village must have vast quantities of it. The water requirements for drinking and washing and cooking were modest. The requirements for the village's thousands of horses were not. They were high. The village must be located on a substantial stream, and these were limited in the area. A further limitation was that the stream must not be navigable by steamboats. Otherwise, the village would be detected quickly and resources could be brought to bear to ensure its destruction. These limitations greatly limited the search area.

The prime search area was four tributaries of the Yellowstone River and their tributaries. The four tributaries were, from east to west, the Powder, the Tongue, the Rosebud, and the Big Horn. The four all ran roughly from south to north, were roughly parallel, were about 15 to 20 miles apart, and were separated by high ridges (see map 34).

It was within a box extending from the Yellowstone to a line about 100 miles to the south, and from the Powder in the east to the Big Horn in the west that the hostiles were expected to be found. This, however, was still an area encompassing 10,000 square miles. Inasmuch as the village must be on the water, the search area was compressed to probably no more than 1,000 lineal miles.

It might seem a large undertaking to examine 1,000 miles of stream front, but it was really not that difficult. High ridges ran between the tributaries of the Yellowstone, and the village had to be at the bottom, at the water's edge. Thus, it could be seen at a distance, and the smoke from it at a greater distance.

A second vulnerability was that the Indians periodically moved their village, and this left a trail. Upon the mention of a trail, one often thinks of something such as a snapped twig that would only be noticed by the most experienced tracker. The trail left by the moving Indians was nothing of the sort. It was such that it could be detected easily by the

Map 34
Custer's Last Stand

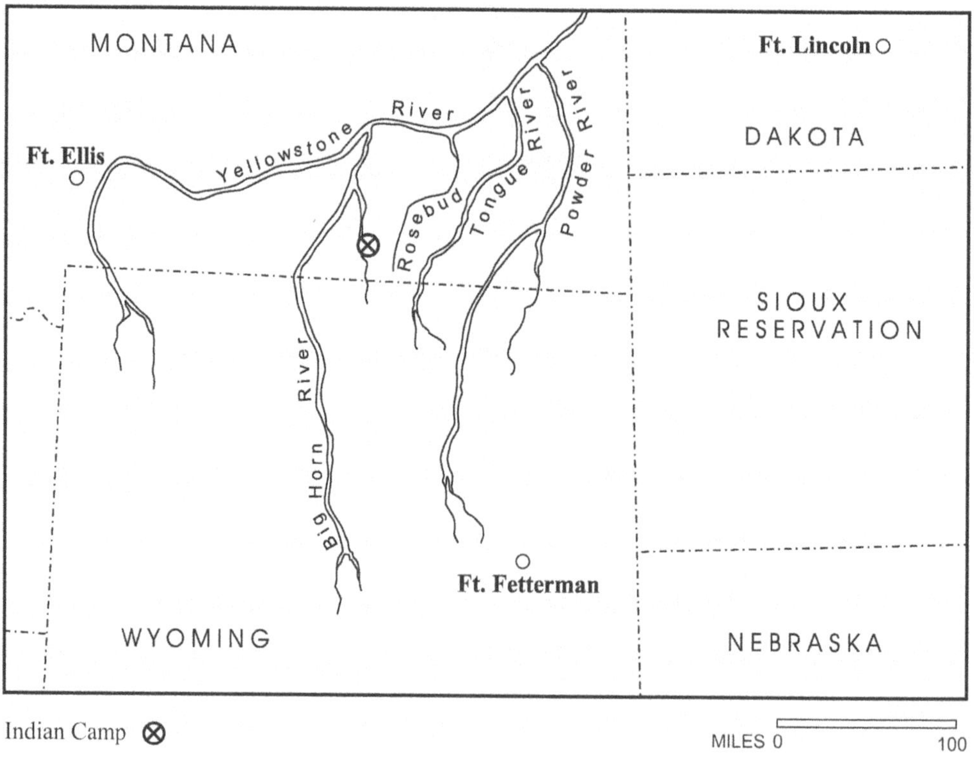

blind. The Lakota Indians transported their goods by travois. A travois consisted of two poles. One end of each pole was tied to a side of a horse. The other ends of the poles bounced along the ground as the horse advanced. The goods of the Indians were placed on a platform between the poles. The bouncing poles scarred the ground, as did the horses' hoofs. A trail that involved hundreds of travois and thousands of horses would leave a scarred ground perhaps one-half mile wide, liberally sprinkled with horse manure. Periodically, the trail would be punctuated with the remains of hundreds of campfires where the Indians stopped for the night.

The Army Plan

The army planned a winter campaign for the winter of 1875-76 and a following spring-summer campaign in 1876. The winter campaign came to nothing as a result of the ferocious weather in the area. It all came down to the spring-summer campaign.

The army would operate from three bases on the periphery of the Indian territory. These were Fort Lincoln in the east, Fort Fetterman in the south, and Fort Ellis in the west (see map 34). The distance between the forts was vast. It was almost 500 miles from Ellis to Lincoln, and about 300 miles from Ellis and Lincoln to Fetterman.

A column was to proceed from each of the three forts so as to arrive in the Yellowstone tributary area in mid–June. The column from Fort Ellis would be under the command of Colonel John Gibbon. It would consist of six companies of infantry and three of cavalry — a total of 450 in all. It would march eastward down the north bank of the Yellowstone River until it met the column from Fort Lincoln in the vicinity of the mouth of the Powder River.

The column from Fort Lincoln was under the command of General Terry. Terry's column included the entire Seventh Calvary under Lieutenant Colonel Custer, two companies of infantry, a Gatling gun detachment, 150 mule-pulled wagons containing supplies, and a large detachment of pack mules. This force was to be joined by three companies of infantry that were to be transported by river, up the Missouri and down the Yellowstone, until they joined in the area of the Powder River. The three companies of infantry were to be transported by the steamer *Far West*, which was to remain and serve as General Terry's headquarters. Terry's total force numbered 1,049. After the forces of Terry and Gibbon met on the Yellowstone, Terry was to assume command of the whole.

Thus, by late June, the U.S. would have a force composed of about 1,500 under Terry assembled on the Yellowstone ready to proceed south down the Yellowstone tributaries against the presumed location of the Indians. At the same time, a similar sized force under General Crook would be approaching from the south and, once the Indians were caught in the closing vise, voila, it would all be over.

Unfortunately, there were a few little problems with the plan. For one, it was predicated upon the assumption that there were not more than 500 to 800 hostile Indians. In this case, had Gibbon or Crook or Terry happened to have to fight the Indians alone, they could have prevailed. The assumption as to the number of hostiles proved to be wildly wrong. For weeks before the plan was implemented, agents of the Department of the Interior noted that Indians were leaving the reservations in droves to the join the hostiles. As was typical, the Department of the Interior failed to advise the War Department of this vital information until weeks after the campaign came to a disastrous end. The actual number of hostiles was estimated by General Sheridan as 2,500 to 3,000[2] and by General Fry as 2,500.[3] Other estimates ranged as high as 4,000 warriors. Although there is no consensus as to their exact number, it appears there were not less than three times the number on which the plan was based.

Another little problem was the fact that, when the three columns set out, they did not know where the Indians were. They presumed they were in the Yellowstone tributary area, which they were, but this was still a vast area.

Lastly, we have the problem of communications. There were absolutely no communications between Terry and Gibbon until they met, and absolutely no communications between Terry and Crook until after the campaign was over. The nearest telegraph station to the mouth of the Powder River was 500 miles.

Execution of the Plan

The three columns set out in the spring hoping to reach their objectives by mid–June: Gibbon down the Yellowstone to meet Terry; Terry up the Yellowstone to meet Gibbon and consolidate their forces to then march south against the Indians; and Crook to approach the Indians from the south.

Terry reached his objective, the mouth of the Powder, first. He arrived by the end of May. Here he established a forward supply base. He parked his 150 wagons and pastured the beef cattle that he had brought along. The wagons contained 30 days of forage, 30 days of rations (minus beef, which was on the hoof), and extra ammunition. The steamship *Far West* arrived and disembarked 300 troops that were to serve as the garrison of the forward base, and Terry then established the ship as his headquarters.

While waiting for Gibbon to arrive, Terry mounted a reconnaissance to pinpoint the location of the Indians. On June 10, he ordered Major Reno of the Seventh Cavalry to take six companies of the regiment (about 300 men) and proceed down the Powder River examining its tributaries, and then to cross over to the Tongue River and come back up to the Yellowstone, examining the Tongue tributaries en route.

Reno departed on June 11 with 10 days of rations. Upon Reno's departure, Terry ordered the *Far West*, with himself and staff embarked, to proceed up the Yellowstone to the mouth of the Tongue to await Reno's return. He also ordered Custer, with the remaining six companies of the Seventh Cavalry, to march up the Yellowstone from the base at the mouth of the Powder to the mouth of the Tongue, and to reassemble the regiment when Reno returned.

The first report from Reno came in on the 19th, indicating that he had found a large Indian trail. Reno himself returned the next day with full details. He discovered the trail on his way back up the Tongue. The trail led westward from the Tongue to the roughly parallel Rosebud River. It then proceeded south along the Rosebud. The age of the trail was estimated to be three weeks. At each stopping point on the trail, Reno counted the remains of up to 400 campfires so, obviously, a large number of Indians were involved. Reno followed the trail south for 45 miles before giving it up as his rations were running low. Upon hearing Reno's report, it was fairly obvious to Terry where the Indian village must be. A short distance beyond where Reno turned back, a tributary of the Big Horn that ran to the west of the Rosebud, called the Little Big Horn, came close to the Rosebud. Had the Indians intended to establish their village on the Rosebud, they would not have passed several suitable sites. The Little Big Horn provided an ideal site. Terry concluded that it was here they were headed.

On the 21st, the day after Reno returned, Colonel Gibbon reported in to General Terry. Gibbon had left his force bivouacked on the Yellowstone at the mouth of the Big Horn and then rode the last few miles to report to Terry in person. Terry now had his force united and he knew where the Indians were. His supply base was established at the mouth of the Powder, Custer was in camp with his full regiment at the mouth of the Tongue, and Gibbon was in camp at the mouth of the Big Horn. It was now time for action. Terry called for a meeting on the *Far West* that very night. The meeting was to include Custer, Gibbon, and Major Brisbin, the commander of Colonel Gibbon's cavalry detachment.

As Terry's group sat down for their meeting, there was one vital piece of information they did not know. The overall plan called for General Crook to be advancing on the hostiles from the south as General Terry advanced upon them with a similar sized force from the north. As of mid-June, Crook indeed was coming up from the south and was on the Rosebud. Crook did not know where the hostiles were, but had a large contingent of 260 Crow and Shoshone Indian scouts who had some familiarity with the area to help him find them. As of June 17, Crook was some 30 miles south of the Indian village on the Little Big Horn, while Reno was at his southernmost and turn-around point just ten miles or so north of the Indian village. Although the two commands were now no more than 40 miles apart, each knew nothing of the other's presence. At this point, Reno knew pretty well where the Indian village was, but Crook did not.

On June 17, some of the hostiles stumbled upon Crook, and warriors by the hundreds rode out from the village on the Little Big Horn to meet him. This resulted in the "Battle of the Rosebud." Over 1,000 Indians ultimately confronted Crook, and the two sides each suffered 30 to 60 casualties. Although Crook successfully fought off the Indian attacks, he was surprised by the numbers confronting him and the ferocity of the attacks.

Considering that Crook was now low on supplies and ammunition, and considering the surprisingly large number of Indians encountered, Crook selected caution over valor and retired to his supply base to the south to await reinforcements. He was now effectively out of the campaign. Inasmuch as there were no communications between Crook and Terry, Terry knew none of this as he sat down to his meeting on the *Far West*.

In a way, Terry had lucked out. Had the hostiles stumbled upon Reno with his 300 rather than on Crook and his 1,500, Reno very likely would have been wiped out.

At the meeting, Terry outlined his plans for his subordinates. The presumed location of the Indian village was on the Little Big Horn. It was roughly 90 miles south of the Yellowstone. Gibbon was camped at the mouth of the Big Horn. He could march directly down the Big Horn to its tributary, the Little Big Horn, and then follow the Little Big Horn directly to the village. This was the most direct route. Custer, with the Seventh Cavalry, was at the mouth of the Tongue. He could follow the Tongue down to the Indian trail that crossed over to the Rosebud and then follow the Rosebud down to the village. This was slightly longer. Alternatively, Custer could march up the Yellowstone from the mouth of the Tongue to the mouth of the Rosebud and then proceed down the Rosebud until he picked up the Indian trail.

At this point, let us turn to Terry and see, in his own words, what he outlined as the plan during the meeting. Gibbon was to proceed south up the Big Horn and Custer was to proceed south up the Rosebud and ascertain the direction of the Indian trail and:

> ... if it led to the Little Big Horn, it should not be followed, but that Custer should keep still further to the south, before turning to the river, in order to intercept the Indians should they attempt to pass around to his left, and in order by a longer march, to give time for Gibbon's column to come up.... This plan was founded on the belief that the two columns might be brought into cooperating distance of each other, so either of them which should be first engaged might by a "waiting fight" give time for the other to come up.[4]

General Terry's intention is thus abundantly clear. Reno had discovered the trail of the hostiles. He had followed it for 45 miles straight down the Rosebud and then returned to base. Shortly farther down the Rosebud, a tributary of the Big Horn that was 15 miles to the west of the Rosebud or roughly parallel to it ran close to the Rosebud. The name of the tributary was the Little Big Horn, and it was here that Terry and his staff assumed the hostiles would make their camp. He was now directing Custer with his cavalry regiment to proceed down the Rosebud, and Gibbon with his mixed infantry and cavalry to proceed down the Big Horn. Inasmuch as Custer had all cavalry and Gibbon had both infantry and cavalry, it was correctly assumed that Custer would arrive in the area of the Indians first. Consequently, Terry intended that Custer not turn off the trail where the Indians turned off but continue farther south and then turn off to his right. This would place Custer south of the Indians and also give Gibbon's force additional time to arrive and engage the Indians from the north. Inasmuch as the presumed Indian camp was 90 miles south of Custer's and Gibbon's start points, and both started out early on the 22nd, Gibbon could not arrive at the Indian camp before the 26th. It was assumed that the combined forces would attack the Indians that date: Gibbon from the north, Custer from the south.

At the conclusion of the meeting, Terry gave Custer his orders in writing. Once they parted, they were never to contact each other or meet again. What would govern Custer's movements was what was on that piece of paper. Unfortunately, there are two versions of the piece of paper and, although they agree in substance, they vary in wording and emphasis. We will first present the version that Custer most likely carried with him, which was from Terry's files:

Camp at the Mouth of the Rosebud River,
Montana Territory, June 22, 1867
Lieut. Col. Custer, 7th Cavalry
COLONEL:

The Brigadier-General Commanding directs that, as soon as your regiment can be made ready for the march, you will proceed up the Rosebud in pursuit of the Indians whose trail was discovered by Major Reno a few days since. It is, of course, impossible to give you any definite instructions in regard to this movement, and were it not impossible to do so, the Department Commander places too much confidence in your zeal, energy and ability to wish to impose upon you precise orders which might hamper your action when nearly in contact with the enemy. He will, however, indicate to you his own views of what your actions should be, and he desires that you should conform to them unless you shall see sufficient reason for departing from them. He thinks that you should proceed up the Rosebud until you ascertain definitely the direction in which the trail above spoken of leads. Should it be found (as it appears almost certain that it will be found) to turn towards the Little Horn, he thinks that you should still proceed southwards, perhaps as far as the headwaters of the Tongue, and then turn towards the Little Horn, feeling constantly, however to your left, so as to preclude the possibility of the escape of the Indians to the south or southeast by passing around your left flank. The column of Colonel Gibbon is now in motion for the mouth of the Big Horn. As soon as it reaches that point it will cross the Yellowstone and move up at least as far as the forks of the Big and Little Horns. Of course its future movements must be controlled by circumstances as they arise, but it is hoped that the Indians, if upon the Little Horn, may be so nearly enclosed by the two columns that their escape will be impossible.

The Department Commander desires that on your way up the Rosebud you should thoroughly examine the upper part of Tullocks' Creek and that you should endeavor to send a scout through to Col. Gibbon's column, with information of the result of your examination. The lower part of the creek will be examined by a detachment from Colonel Gibbon's command. The supply steamer will be pushed up the Big Horn as far as the forks of the river if found to be manageable for that distance, and the Department Commander, who will accompany the column of Colonel Gibbon, desires you to report to him there not later than the expiration of the time for which your troops are rationed, unless in the meantime you receive further orders.

Very respectfully
Your Obedient Servant
E. W. Smith, Captain 18th Infantry
Acting Assistant Adjutant-General[5]

The Tullock's Creek referred to in paragraph two of the order was a small stream between the Rosebud and Big Horn Rivers just north of the Little Big Horn. There were no Indians there, so what Custer may have done or not done in regards to this paragraph has no significance to our story.

The second version of the written order presented to Custer was provided by Major Brisbin, who attended the meeting aboard the *Far West*. After Custer's death, Terry appointed Brisbin commander of all his cavalry, that is, Brisbin's own Second Cavalry Battalion and the remains of Custer's Seventh Cavalry. Upon his appointment, Brisbin states that he looked over all the papers relating to the campaign and took a copy of the order to Custer.

In a letter to General Godfrey dated January 1, 1892, Brisbin alleged that he had the copy before him and that, beginning with the words "You should proceed up the Rosebud," there were some significant differences from the version generally accepted as the one given to Custer. Brisbin's version of the order read as follows:

> ... You should proceed up the Rosebud until you ascertain definitely the direction in which the trail spoken of above leads. Should it be found, as it appears to be almost certain that it will be found, to turn toward the Little Big Horn he thinks [that is, the Department Commander thinks] that you should proceed southward, perhaps as far as the headwaters of the Tongue River and then ("then" was underscored in the order) turn toward Little Big Horn, feeling constantly to your left, so as to preclude the possibility of the escape of the Indians to the south or southeast by passing around your left flank. It is desired that you conform as nearly as possible to these instructions and that you do not depart from them unless you shall see absolute necessity for doing so... [etc., the rest was the same].[6]

The Brisbin version of the order to Custer was much more emphatic as to what Custer was to do than the accepted version. In telling Custer not to turn off when the trail turned off toward the Little Big Horn, it not only told Custer to not turn off, but to continue as far south as the headwaters of the Tongue and "then" turn off—with the word "then" underlined. The Brisbin version further contained the additional caveat that Custer was not to deviate from the instructions unless he saw an "absolute necessity" for doing so.

Before we follow Custer up the Rosebud in the execution of his orders, let us digress to take a look at the Seventh Cavalry Regiment and the leading characters in this unfolding drama.

The Seventh U.S. Calvary

Contrary to popular belief, Lieutenant Colonel Custer was not the commanding officer of the Seventh Cavalry as it marched out on June 22, 1876. The commanding officer was Colonel Sturgis. However, Sturgis was away on temporary duty, and Custer, who was second in command, was present and the de facto commanding officer.

The Seventh, like other regular army cavalry regiments, was broken up into 12 companies, designated A through M. There was no company J. Each company contained about 50 men and was supposed to be commanded by a captain, who had a lieutenant as assistant. However, all of the billets were not filled. Four of the companies were headed by first lieutenants, and three lacked the lieutenant assistants.

In addition to the 12 companies, the regiment had a headquarters detachment, a pack train detachment, and an Indian scouts detachment. The headquarters detachment included Custer; Major Reno, who was second in command; Lieutenant Cooke, the regimental adjutant; and three doctors. The Indian scouts detachment was headed by two officers and contained five guides, two interpreters, and 35 Indian scouts who were paid as enlisted men. The regiment also had one important addition who accompanied the headquarters detachment, Mr. Mark Kelly, reporter for the *New York Times*. The total strength of the regiment as it marched off that morning was 31 officers, 566 enlisted, 15 civilians, and 35 to 40 Indians.[7]

The company commanders of the Seventh Cavalry were not typical of those one normally found in a peacetime regiment. Most were Civil War combat veterans and some had served in much higher ranks during the war, commanding regiments and, in the cases of Major Reno and Captain Benteen, brigades.

Nepotism was rampant in the regiment. Custer's brother, Tom, was commanding officer of Company C, and his brother-in-law, Lieutenant Calhoun, commanded Company L. In addition, Custer's brother, Boston Custer, and his nephew, Henry Reed, were contractors working with the pack train.

None carried a saber that morning. Each enlisted man was armed with a single-action, 45-caliber, six-shot revolver, and a single-shot Springfield 45 caliber "trapdoor" carbine. The pistol had an effective range of only 60 yards, while the carbine's range was at least ten times that. The "trapdoor" carbine received its name because of the way it was loaded. The bolt was hinged and opened toward the muzzle. It fired modern type bullets, but the shell was made of copper rather than brass. Consequently, when fired, it expanded and occasionally stuck and had to be extracted with a pen knife blade. When not jamming, a user could easily get off six or more aimed shots a minute.

The pack mules carried 15 days' rations of hard tack, coffee, and sugar; 12 days' rations of bacon; and 50 rounds of carbine ammunition per man. Each man carried 100 rounds of carbine ammunition and 24 rounds of pistol ammunition, to be carried on his person and in his saddle bags; and each man was to carry on his horse 12 pounds of oats.[8]

General Terry offered Custer the two Gatling guns, but Custer declined to take them because he feared they would slow his progress.

The Leading Characters

General Terry was one of the very few individuals of the Civil War who rose to corps command without being a West Point graduate or receiving political influence. General Grant's assessment of Terry in his personal memoirs says it all. Grant wrote:

> General Alfred H. Terry came into the army as a volunteer without a military education. His way was won without political influence up to an important separate command — the expedition against Fort Fisher, in January, 1865. His success there was most brilliant, and won for him the rank of brigadier-general in the regular army and major-general of volunteers. He is a man who makes friends of those under him by his consideration of their wants and their dues. As a commander, he won their confidence by his coolness in action and by his clearness of perception in taking in the situation under which he was placed at any given time.[9]

Terry's finest hour was yet to be. He lived in an era when many generals put their reputations beyond any cause. They would engage in polemics until their deaths. The disaster at Little Big Horn was clearly Custer's fault and not Terry's, but Terry was often blamed. Terry never attempted to defend himself or denigrate Custer.

Our major character is, of course, George Armstrong Custer. Custer was born on December 5, 1839, and graduated last in his class at West Point in June 1861. But for the war, he may not have graduated at all. Custer saw action continuously in the war from Bull Run to Appomattox. On June 28, 1863, just two years after receiving his commission as second lieutenant, he was promoted directly from captain to brigadier-general. He thus became the youngest general at the time at the age of 23. That was not all. Just two years more, he was promoted to major general, at which rank he ended the war.

It was beyond dispute that Custer was brave and that he made a significant contribution to winning the war. In fact, his contribution was such that General Sheridan presented him with the table on which Lee and Grant had signed the armistice.

Was Custer's meteoric rise then based solely on his outstanding abilities? Any student of the Civil War cannot help but notice the prevalence of photographs of Custer. A good case could be made that he was the most photographed individual of the war. Significantly, he not only appears in the foreground of pictures in his self-designed, gaudy uniforms after he became famous, but in the backgrounds of photographs of other prominent generals before he became famous.

Custer was undoubtedly a rabid seeker of fame and recognition and, just possibly, an apple

Lieutenant Colonel George A. Custer USA: The man who died as he wanted.

polisher par excellence in the bargain. He managed to ingratiate himself with every top general under whom he served. His undoubted bravery may be attributed to his hierarchy of values, wherein he put the desire for fame and recognition above self-preservation.

At the end of the war, when the wartime volunteer army was disbanded, Custer reverted to the rank of captain in the regular army. In 1866, however, he was given the lieutenant colonelcy of the newly formed Seventh Cavalry regiment. He was to remain in this rank and in this position until his death 10 years later.

During the war, although he served in high rank, he never exercised independent command. Now, on the frontier, he was on his own — and quickly got into serious trouble. These were not minor misjudgments but major infractions of the laws, which could have brought him dismissal or even jail. Custer was charged with abandoning his post at Fort Wallace and traveling over 100 miles overland with armed escort to visit his wife at Fort Reilly. If that were not enough, he was charged with ordering some deserters to be shot on sight while, according to the Articles of War, a person could only be executed by conviction by court-martial.

Custer was tried at Ft. Leavenworth in October 1867, and found guilty. Surprisingly, he was only sentenced to be removed from duty for one year without pay. Owing to the intervention of his patron, General Sheridan, he was returned to command in just nine months.

In order to redeem himself, he made a daring attack on an Indian village in November 1868. This resulted in his "victory" in the battle of the Washita, for which he received nationwide publicity, adulation, and fame. However, this victory did have some unsavory loose ends. He claimed to have killed 103 braves.[10]

Other better estimates placed the Indian casualties at about 50 total, killed and wounded, of which most were women and children. Additionally, during the battle, Major Elliott and some 20 troopers chased off after some retreating Indians. Custer left the battle area for home without ever checking to see what happened to Elliott. Elliott and his entire group were later found massacred.

By the time Custer set out down the Rosebud on June 22, 1876, he had again managed to establish himself in the minds of the public as George Armstrong Custer, beau ideal, war hero, and Indian fighter supreme.

Custer's second in command, who was to play a major part in the unfolding drama, was Major Marcus Reno. Reno was 42 in 1876, five years older than Custer, and had graduated from West Point in 1857, four years before Custer. Unlike Custer, who graduated at the bottom of his class, Reno graduated near the middle of his, 20 out of 38. Reno had a successful Civil War career. He saw much combat and was brevetted twice, once for gallantry and once for meritorious conduct. He ended the war as a brevet brigadier-general in charge of a brigade.

At the end of the war, Reno reverted to the rank of captain in the regular army. He was promoted to major in 1868 and, in December of that year, assigned to the Seventh Cavalry. From that date until Custer's death in 1876, Reno's fate was tied to that of Custer. Reno was next in command.

What were Reno's true feelings regarding Custer? In the court of inquiry conducted in 1879, this question was put to Reno and Reno had to answer under oath. Reno replied that he had no animosity and that he and the general got on well enough.[11] The questioner persisted. Reno responded, "My feelings toward General Custer were friendly."[12] The questioner persisted further and Reno replied, "Well sir, I had known General Custer a long time, and I had no confidence in his ability as a soldier."[13]

The third in the chain of command at the time of the Little Big Horn was Captain Frederick Benteen. Benteen, like Reno, was 42 years old. Unlike Reno and Custer, Benteen did not attend West Point. In fact, at the outset of the war, he had no military experience.

Benteen came from a family of rabid secessionists with a southern heritage. When he announced his intention to serve the Union and accepted a lieutenancy in the Tenth Missouri, his father disinherited him and, according to Benteen, said, "I hope the first god damned bullet gets you."[14]

Benteen participated in actions in the west, large and small, until the end of the war. At war's end, he had risen to the rank of colonel and was recommended for the rank of brevet brigadier-general. During all his wartime service, his subsequent service against the Indians before and after Little Big Horn, and at the Battle of Little Big Horn, Benteen displayed an unusual degree of calm bravery that inspired others. Lieutenant Varnum later remarked that Benteen (at the battle of Little Big Horn) was the only man he ever saw who did not try to dodge bullets. Benteen seemed oblivious to danger; he walked around checking on his troops, deliberatively inviting fire, but was hurt just once when a bullet nicked his thumb.[15] Benteen himself, when later describing the episode, modestly remarked, "I state but the facts when I say that we had a fairly warm time with those red men."[16]

Benteen, unlike Custer, was not vainglorious and did not solicit fame and publicity. In fact, he was what one might call a private person. Nevertheless, he was generally well-liked by both his fellow officers and his subordinates.

After the Civil War, when the U.S. Volunteer Army was disbanded, Benteen received an appointment as captain of the newly formed Seventh Cavalry regiment, now headed by Custer. Benteen reported to Custer for duty in January 1867. His fate was now tied to that of Custer until the death of Custer in 1876. From Benteen's initial meeting with Custer, he did not like him, considering him a braggart.

Benteen's dislike of Custer was intensified by Custer's conduct at the battle of the Washita. He considered Custer's abandoning the field without looking for Major Elliott and his force a disgrace. Elliott and his force were later found dead and mutilated. Whether or not Custer could have saved them is, of course, not known. In Benteen's eyes, it was inexcusable for him not to try.

Benteen was never intimidated by Custer, and Custer probably knew instinctively that he could push Benteen only so far. In February 1869, Benteen wrote a personal letter to a friend in St. Louis that was highly critical of Custer's conduct at the battle of the Washita. The friend provided its contents to the *St. Louis Democrat*, without Benteen's knowledge or permission. The *Democrat* published it and it was picked up by the *New York Times* and became national news. The article came to Custer's attention and he was furious.

Custer sounded officers call. He displayed the newspaper to the assembled group and announced that, if he found out who was responsible, he meant to cowhide that party. Benteen asked to see the paper. After reading a few lines, he stepped outside and twirled the cylinder of his revolver. He then stepped back inside and said, "I guess I am the man you are after, and I am ready for the whipping promised." Custer turned red and said, "Colonel Benteen, I'll see you again, sir." He then dismissed the officers and the matter came to an end.[17]

We now come to June 1876, without any further known major conflict between Custer and Benteen.

The Execution of Terry's Plan

Now let us proceed with a description of the execution of Custer's part of Terry's plan. To review: The Indian village was believed to be on the Little Big Horn, about 90 miles south of Terry's position on the Yellowstone River. The rivers Rosebud and Big Horn ran south to north, were roughly parallel and about 15 miles apart, with the Little Big Horn in between. The plan envisioned two columns enveloping the Indians and then cooperating in their defeat. Custer was to proceed down the Rosebud and Gibbons, accompanied by Terry, was to proceed down the Big Horn. The steamer, *Far West*, was to proceed down the Big Horn as far as it could. Custer's column was all cavalry, and Gibbon's both infantry and cavalry. Consequently, Custer was expected to arrive in the area of the Indians well before Gibbon. The plan called for Custer not to attack the Indians, but to continue on south past the village and then take a position south of the Indians. This was to give the slower Gibbon time to arrive at the north end of the village before the attack commenced. The distances were such that Gibbon could not possibly arrive at the Indian camp before the 26th, and would probably not arrive until the 27th. The plan required Custer to report to Terry aboard the *Far West* "not later than the expiration of the time for which your troops are rationed," i.e., 15 days, on July 6. With this background, let us follow Custer's movements day by day.

June 22

Custer received the plan in writing the night of June 21–22. The regiment departed its camp on the Yellowstone at the mouth of the Rosebud at noon on the 22nd. Trouble with the pack mules started almost immediately. Packs fell off and mules straggled. After a march of only 12 miles, the regiment went into camp at 4 P.M. Custer issued an officers call at dark. He decreed that the next day's march would start at 5 A.M., each day's march from this point onward would be 25 to 30 miles, and that there would be no use of the bugle beyond this point. He estimated the number of warriors that they would encounter to be as many as 1,500, but no more.

June 23

The regiment started at 5 A.M. sharp. After eight miles, they picked up the Indian trail and passed through the first of the vacated Indian camping places. During the day, they passed through three additional Indian camping places. It was abundantly clear from the size of the camps that the number of Indians was large. The regiment covered 33 miles during the day and went into camp at 5 P.M.

June 24

On June 24, the regiment again set out at 5 A.M. and, throughout the day, passed through multiple vacated Indian camp sites. At this time, they assumed that each site was a successive site of the same group of Indians. This proved to be erroneous. Several adjacent sites were camp grounds of different tribes of Indians in the same encampment. Thus, the group of Indians they were following was far greater than they imagined.

The regiment camped at dark at a large vacated Indian camp site. It had covered 28 miles that day. Custer issued an officers call. It was already dark and officers groped in the dark toward the light of a single candle in Custer's tent. Custer announced: The trail now moved to the right over the divide to the Little Big Horn. His Crow scouts had discovered fresh tracks of Indians in the vicinity. The command would be ready for a night march by 11:30 P.M. He was anxious to get close to the divide between the Rosebud and Big Horn before daylight. They would spend the day of the 25th reconnoitering the Indian encampment and attack on the morning of the 26th. All fires would be extinguished after supper tonight, and there would be none until after the attack.

June 25

The regiment marched 10 miles during the night of the 24th–25th and halted at the divide between the Rosebud and Big Horn at 2 A.M. to await further information from their Indian scouts. About 8 A.M., General Custer rode among the resting troops and announced that they had been discovered and that they would march to the attack immediately. In fact, he was in error. They had not been discovered. During the night march, a carton containing bread had fallen off one of the pack mules. A trooper sent back sighted some Indians recovering the carton and moving off and he so reported. Unknown to Custer, the Indians recovering the carton were not part of those encamped and did not report the presence of the soldiers.

As the tired troopers set out that morning, the Indian village was still some 15 miles away to the northwest. As Custer's regiment trotted toward the village on the Little Big Horn on the morning of June 25, 1876, Custer smelled blood and was a dynamo of activity. His main concern appeared to be that the Indians would get away from him and all his actions were based on that premise.

The battle of Little Big Horn has perhaps been reported upon and dissected by historians more than any other battle. For our purposes, we need to present only a general overview. As Custer approached the village, he began to divide his command. First, he ordered Benteen to take three of his 12 companies and proceed at right angles off the trail, apparently to see that there was no way for the Indians to retreat to the south, or to ascertain that there were no more Indians south of the village. He next ordered one of the remaining nine companies to proceed to the rear and accompany the lagging pack train. He then divided the remaining eight companies into two battalions, five directly commanded by him, and three by Major Reno. The two battalions continued to trot side by side toward the village with Reno on the left and Custer on the right. When the southern end of the village came into view, down below the bluff on which they were riding and on the far side of the Little Big Horn River, Custer ordered Reno to proceed down the bluff, cross the river, and attack the village. Custer and his battalion continued northward along the ridge, sometimes in sight of Reno below and sometimes not.

Reno deployed his men in line and began a charge toward the village. However, he quickly realized the magnitude of the force gathering to confront him and ordered his men to dismount and proceed on foot.

As Reno's battalion proceeded on foot toward the village, his left flank was in the air and he could see clouds of Indians riding around the end of the line to attack him from the rear. He ordered a retreat to a woods on the village side of the river, where he formed a defense line with his back to the river. By this time, he could no longer see Custer proceeding along the ridge above — and was never to see him again.

Reno soon concluded that his position in the woods was untenable. A contributing factor was the death of his chief Indian scout, Bloody Knife. While he was talking to Bloody Knife, a bullet struck Bloody Knife in the head and splashed his brains on Reno's face. A shaken Reno then ordered his men to mount and follow him. The battalion made a wild, disorganized scramble through the Indians, crossed the river, and reached the top of the ridge, where they set up a defense position. By this time, Reno had lost about one-third of his men and had many wounded.

While Reno was still fighting in the woods, Custer reached what he thought was the north end of the village and, probably for the first time, realized the magnitude of the force confronting him. He directed that a courier take a message to Benteen for Benteen to join him. Custer's assistant adjutant general, Lieutenant Cooke, wrote out the message and handed it to the courier. The message read, "Benteen, come on. Big village. Be quick. Bring packs. P. S. Bring Packs."[18]

As the courier rode back along the ridge, the situation was as follows: Reno was still fighting in the woods below and had not yet retired to the ridge. Benteen, having ridden off at right angles to the trail, had soon found that the terrain was such that no movement to the south of the village could be expected, and that there were no Indians south of the village. Consequently, he returned to the trail and was proceeding northward toward the village when the courier found him. The pack train was still far behind Benteen. When the courier handed Benteen the message, Benteen was probably about 10 miles from Custer.

As Benteen proceeded northward along the ridge, he met, not Custer, but a distraught Reno trying to organize a defense. What exact words transpired between Reno and Benteen we do not know. However, if Reno told Benteen to help him, that is what Benteen was required to do. The so-called "rule of good discipline," which every U.S. officer is taught, is that one must obey the last order one receives from competent authority, and not the one issued by the highest authority.

For example, I, as a career naval officer, was taught that if my ship was proceeding on a mission tasked by the highest authority and by chance encountered another U.S. ship senior to me, I had to send the message, "Request authority to proceed on mission assigned." The senior would normally routinely grant the request, but if he did not, I had to obey him.

In any event, Reno and Benteen waited for the pack train to join them. When it did and they finally set out to join Custer, it was too late. Custer's force had probably already been annihilated. Every man of the five companies with Custer was killed. The losses included Custer's brothers Tom and Boston, his brother-in-law Lieutenant Calhoun, and his nephew Henry Reed. Reno and Benteen were forced back to the position where they met.

The Reno-Benteen-pack train force spent restless nights of the 25th and 26th surrounded by Indians and nursing their many wounded with the single surviving doctor. Late on the 26th, they noted the Indians packing up and leaving. On the morning of the 27th, the Terry-Gibbon column arrived and they were saved.

An Analysis

Two battlefield orders impacted on events. First, let us look at the order from Custer to Benteen, and then at the one from Terry to Custer.

The order from Custer to Benteen ordered "Come on," "Be quick," and "Bring Packs." "Bring Packs" alone was repeated, attesting to its importance. When one orders someone to "bring" something, it would normally mean to the average person that the thing to be brought would be in the possession of the person ordered to bring it. This is precisely what Reno and Benteen did. They did not proceed to join Custer until the packs arrived at their location and were then in their possession. By then, however, it was too late. A court of inquiry was convened after the battle and both Reno and Benteen were exonerated, although not commended.

Supposing the order simply read, "Benteen come on. Big village. Be quick" and ended there; or read, "Benteen come on. Big village. Be quick. Expedite packs." Had either of these wordings been used, it just might have turned out differently.

Now let us turn to the order from Terry to Custer. It is absolutely clear that Custer's actions did not conform to Terry's intentions. Terry's intentions expressed in both the version of the order believed to have been given to Custer and Major Brisbin's version were that, "You should proceed up the Rosebud until you ascertain the direction in which the trail above spoken leads. Should it be found, as it appears to be almost certain that it will be found, to turn toward the Little Big Horn, that you should still proceed southward, perhaps as far as the headquarters of the Tongue River and then turn toward the Little Big Horn."

Both versions of the order were thus in agreement that Custer was not to follow the trail directly to the Indian village, but to continue on down south past the turn-off as far as the headwaters of the Tongue and then turn off in the direction of the Little Big Horn. The Brisbin version of the order even underlined the word "then" to emphasize the point. This Custer did not do, but rather, followed the trail directly to the Indian village and attacked it.

The two versions of the order also varied in the degree of discretion they gave Custer. The version believed given to Custer read, "He [Terry] will however, indicate to you his own views of what your action should be, and he believes that you should conform to them unless you see sufficient reason for departing from them." Brisbin's version of the discretion read, "It is desired that you conform as nearly as possible to these instructions and that you do not depart from them unless you shall see absolute necessity for doing so."

Suppose Custer had received the Brisbin version of the discretion allowed rather than the one he apparently did. Would he then have violated the intent of the orders and attacked? Some argue that he would have, that he had strong motivation. Had he proceeded south beyond the village and waited for Gibbon's column, at which Terry was present, to come up before the attack, he would have been third in seniority at the battle, and if the battle had been won, the headlines would have proclaimed "Terry" or perhaps "Terry and Gibbon Crush Hostiles." If he attacked before they arrived and won, as he expected, the headlines would proclaim "Custer Crushes Hostiles."

Custer, however, had an extremely good reason for not directly violating any orders. He had previously been court-martialed and convicted back in 1867 and the man ordering the court-martial and approving the sentence was General Grant. General Grant was now president and Custer's commander-in-chief, and Grant held Custer's fate in the palm of his hand. Shortly before the Lakota War, Custer had again succeeded in offending Grant. He had testified against Grant's brother at a congressional hearing. He had so offended Grant that Grant had ordered that Custer not participate in the current expedition. Grant only reluctantly relented at Sheridan's intercession. Had Custer directly violated his orders and either lost the battle or merely incurred excessively high casualties, and had he been bound by Brisbin's version of the discretion, he would most likely have been court-martialed. Thus, again we have a case where the difference of wording in an order having the same general intent may have determined the outcome of a battle. Had Custer been bound by Brisbin's version of the order, he might not have attacked on the 25th, but awaited the arrival of the Gibbon force as the plan intended. Had he done so. there would have been no massacre at the Little Big Horn, and no "Custer's Last Stand."

Chapter Notes

Abbreviations

OR—U.S. War Department, *The War of the Rebellion: A Compilation of the Official Records of the Union and Confederate Armies*, Series I.
BL—*Battles and Leaders of the Civil War*.

Chapter 1

1. OR, Vol. 5, Part I, 557–558.
2. Ibid., 290.
3. Ibid.
4. Ibid., 291.
5. Ibid., 299.
6. Catton, *The Civil War*, 160.
7. OR, Vol. 5, Part I, 303.
8. Ibid., 301.
9. Ibid., 300.
10. Ibid., 353.
11. Ibid., 308.

Chapter 2

1. OR, Vol. 12, Part 1, 524–25.
2. Ibid., 702.
3. Ibid., 704.
4. Ibid.
5. Ibid., 706.
6. Ibid.
7. Ibid., 709–10.

Chapter 3

1. McClellan, "The Peninsular Campaign," 173.
2. Ibid.
3. OR, Vol. 11, Part III, 589.
4. Hill, "Lee Attacks North of the Chickahominy," 347.
5. Ibid.
6. Ibid.
7. OR, Vol. 11, Part II, 498–499.
8. Ibid., 881–882.
9. Ibid., 835.
10. Ibid.

11. Ibid., 514.
12. Ibid., 499.
13. Ibid., 562.
14. Ibid., 614.
15. Compiled from Official Records by Marcus J. Wright, Chief of the Division of Confederate Records, U.S. War Dept., Francis Trevelyan Miller, *The Armies and the Leaders*, 142.
16. Ibid.

Chapter 4

1. *BL*, Vol. II, 391.
2. Ibid., 388.
3. Ibid., 388–89.
4. Ibid., 402.
5. Ibid., 390.
6. OR, Vol. 11, Part III, 280.
7. Hill, "McClellan's Change of Base and Malvern Hill," 392.
8. *BL*, Vol. II, 393.
9. Ibid., 394.
10. Scharf, "History of the Confederate States Navy," 506.
11. Ibid., 214.

Chapter 5

1. OR, Vol. 11, Part II, 625.
2. *BL*, Vol. II, 500.
3. OR, Vol. 12, Part III, 729.
4. OR, Vol. 12, Part II, Supplement, 825.
5. OR, Vol. 12, Part II, 519.
6. OR, Vol. 12, Part II, Supplement, 955.
7. OR, Vol. 12, Part II, 525.
8. Ibid., 529.
9. OR, Vol. 12, Part III, 735.
10. Ibid., 755–756.
11. OR, Vol. 12, Part II, Supplement, 835.

Chapter 6

1. Wikipedia, http://en.wikipedia.org/wiki/Special_Order_191.
2. OR, Vol. 19, Part I, 818.
3. OR, Vol. 19, Part II, 606.
4. OR, Vol. 19, Part I, 45.
5. OR, Vol. 19, Part II, 294.
6. OR, Vol. 19, Part I, 787.
7. OR, Vol. 19, Part II, 603.
8. OR, Vol. 19, Part I, 959.

Chapter 7

1. *Don Carlos Buell Source Page*, www.aotc.net/Buell_home.htm.
2. Ibid.
3. Hurst, *Nathan Bedford Forrest*, 139.
4. *Leonidas Polk: Southern Civil War General*, www.historynet.com/?s=leonidas+polk.
5. Wheeler, "Bragg's Invasion of Kentucky," 15.
6. OR, Vol. 16, Part I, 1096.
7. Miller, *The Armies and the Leaders*, 142.

8. Snow, *Lee and His Generals*, 437.
9. OR, Vol. 16, Part II, 897.
10. Ibid., 901.
11. OR, Vol. 16, Part I, 1102.

Chapter 8

1. Swinton, "Army of the Potomac," 650.
2. Ibid., 227.
3. OR, Vol. 51, Part I Supplement, 1021.
4. Ibid.
5. "The Battle of Fredericksburg, An Authentic Statement," *New York Times*, March 20, 1863.
6. OR, Vol. 21, 118.
7. OR, Vol. 21, 71.
8. "The Conduct of the War. Report of the Congressional Committee," *New York Times*, April 26, 1863.
9. OR, Vol. 51, Part I Supplement, 1030.
10. "The Conduct of the War. Report of the Congressional Committee," *New York Times*, April 26, 1863 (from testimony of Franklin).
11. Miller, *The Armies and the Leaders*, 142.
12. Oxford Dictionary of Civil War Quotations, 333.
13. "The Conduct of the War. Report of the Congressional Committee," *New York Times*, April 26, 1863 (from testimony of Gen. Burnside).
14. "The Conduct of the War. Report of the Congressional Committee," *New York Times*, April 26, 1863.
15. Sandburg, *Abraham Lincoln: The War Years*, Vol. I, 622.
16. Ibid., 624.
17. Grant, *The Personal Memoirs of Ulysses S. Grant*, 657.
18. Sandburg, *Abraham Lincoln: The War Years*, Vol. I, 154.

Chapter 9

1. Johnston, "Jefferson Davis and the Mississippi Campaign," 473.
2. OR, Vol. 24, Part III, 815.
3. Ibid., 260–360.
4. Johnston, "Jefferson Davis and the Mississippi Campaign," 475.
5. OR, Vol. 24, Part III, 870.
6. Johnston, "Jefferson Davis and the Mississippi Campaign," 479.
7. Ibid.
8. OR, Vol. 24, Part III, 876.
9. Johnston, "Jefferson Davis and the Mississippi Campaign," 479.
10. OR, Vol. 24, Part III, 878.
11. Ibid., 882.
12. OR, Vol. 24, Part I, 241.
13. Editors of the Civil War Times Illustrated, *Great Battles of the Civil War*, 336.
14. OR, Vol. 24, Part I, 241.
15. OR, Vol. 24, Part I, 272–73.
16. Ibid.

Chapter 10

1. Taylor, *Destruction Reconstruction*, 35.
2. McIntosh, *Review of the Gettysburg Campaign*, 52.
3. OR, Series 1, Vol. 27, Part 1, 266.
4. OR, Series 1, Vol. 27, Part 1, 284.
5. Tucker, *High Tide at Gettysburg*, 176.

6. Trimble, *Confederate Veteran*, Southern Historical Society Papers, 273.
7. Taylor, *Four Years with General Lee*, 95.
8. OR, Series 1, Vol. 27, Part 1, 284.
9. Thomason, *Jeb Stuart*, 422–23.

Chapter 11

1. *New York Times*, October 30, 1862.
2. Katcher, *American Civil War Commanders (4), Confederate Leaders in the West*, 25.
3. OR, Vol. 30, Part II, 28.
4. Ibid.
5. Ibid.
6. Ibid, 301.
7. Ibid.
8. OR, Vol. 30, Part II, 294.
9. Ibid., 295.
10. Ibid., 30.
11. Ibid.
12. Ibid.
13. Ibid., 103.
14. Ibid., 59.
15. "Wood at Chickamauga," *New York Times*, December 25, 1881.

Chapter 12

1. Dyer, *The Gallant Hood*, 244.
2. From an address to the West Point graduation class of 1879 by Lt. Gen. J. M. Schofield, Wikipedia, "John Schofield."
3. OR, Vol. 45, Part II, 656.
4. Dyer, *The Gallant Hood*, 288.
5. Ibid.
6. OR, Vol. 45, Part II, 653.
7. Cheatham, "General Cheatham at Spring Hill," 439.
8. Ibid.

Chapter 13

1. OR, Vol. 46, Part I, 234.
2. Porter, "Five Forks and the Pursuit of Lee," *BL*, Vol. IV, 708.
3. OR, Vol. 46, Part I, 796.
4. Ibid., 1110.
5. Ibid.
6. Ibid., 826.
7. *Oxford Dictionary of Civil War Quotations*, 333.
8. Wikipedia, http://en.wikipedia.org/wiki/Philip_Sheridan.
9. Freeman, *Lee's Lieutenants*, 778.
10. OR, Vol. 46, Part I, 828.
11. Ibid., 1105.
12. Grant, *The Personal Memoirs of Ulysses S. Grant*, 604.

Appendix

1. Graham, *The Custer Myth*, 151 (Godfrey's Narrative, *Century Magazine*, January 1892).
2. Ibid.
3. Ibid.

4. Graham, *The Custer Myth*, 152 (General Terry's Annual Report in 1876).
5. Graham, *The Custer Myth*, 133–34 (General Terry's Files).
6. Graham, *The Custer Myth*, 155–56 (Brisbin's letter to Godfry on January 1, 1892).
7. Brininstock, *Troopers with Custer*, 60–62.
8. Graham, *The Custer Myth*, 130 (General Godfrey's Narrative).
9. Grant, *The Personal Memoirs of Ulysses S. Grant*, 658.
10. "Custer's Report to Sheridan of Nov. 28, 1868," U.S. Senate Document 40th Congress, 3rd Session, 1869, Executive Document 13.
11. Connell, *Son of the Morning Star*, 10–11.
12. Ibid.
13. Ibid.
14. Ibid., 31.
15. Ibid.
16. Ibid.
17. Ibid., 197.
18. Graham, *The Custer Myth*, 140.

Bibliography

"The Battle of Fredericksburg, An Authentic Statement." *New York Times*, March 20, 1863.
Battles and Leaders of the Civil War, Vol. II. New York: Thomas Yoseloff, 1956.
Brininstock, E. A. *Troopers with Custer*. Mechanicsburg, PA: Stackpole Books, 1994.
Catton, Bruce. *The Civil War*. New York: American Heritage Press, 1971.
Cheatham, Maj. Gen. B. F. "General Cheatham at Spring Hill." *Battles and Leaders of the Civil War, Vol. IV*. New York: Thomas Yoseloff, 1956.
Civil War Times Illustrated Editors. *Great Battles of the Civil War*. New York: Gallery Books, 1967.
"The Conduct of the War. Report of the Congressional Committee." *New York Times*, April 26, 1863.
Connell, Evan S. *Son of the Morning Star*. New York: Promontory Press, 1993.
"Custer's Report to Sheridan of Nov. 28, 1868." U.S. Senate Document, 40th Congress, 3rd Session, Executive Document 13, 1869.
Don Carlos Buell Source Page. http://www.aotc.net/Buell_home.htm.
Dyer, John P. *The Gallant Hood*. New York: Konecky and Konecky, 1950.
Freeman, Douglas Southall. *Lee's Lieutenants*. New York: Scribner, 1998.
Graham, Col. W. A. *The Custer Myth*. New York: Bonanza Books, 1953.
Grant, Ulysses S. *The Personal Memoirs of Ulysses S. Grant*. New York: Konecky and Konecky, 1886.
Hill, Gen. Daniel H. "Lee Attacks North of the Chickahominy." *Battles and Leaders of the Civil War, Vol. II*. New York: Thomas Yoseloff, 1956.
_____. "McClellan's Change of Base and Malvern Hill." *Battles and Leaders of the Civil War, Vol. II*. New York: Thomas Yoseloff, 1956.
Hurst, Jack. *Nathan Bedford Forrest*. New York: Alfred Knopf, 1993.
"John Schofield." Wikipedia. http://en.wikipedia.org/wiki/John_Schofield.
Johnston, Gen. Joseph. "Jefferson Davis and the Mississippi Campaign." *Battles and Leaders of the Civil War, Vol. III*. New York: Thomas Yoseloff, 1956.
Katcher, Philip. *American Civil War Commanders (4), Confederate Leaders in the West*. Oxford: Osprey, 2003.
Leonidas Polk: Southern Civil War General, www.historynet.com/?s=leonidas+polk.
McClellan, George B. "The Peninsular Campaign." *Battles and Leaders of the Civil War, Vol. II*. New York: Thomas Yoseloff, 1956.
McIntosh, Lt. Col. Gregg. *Review of the Gettysburg Campaign*. Reprint of Old Dominion Edition, 1984.
Miller, Francis Trevelyan. *The Armies and the Leaders*. New York: Castle Books, 1957.
The Oxford Dictionary of Civil War Quotations. New York: Oxford University Press, 2006.
"Philip Sheridan." Wikipedia. http://en.wikipedia.org/wiki/Philip_Sheridan.
Porter, Gen. Horace. "Five Forks and the Pursuit of Lee." *Battles and Leaders of the Civil War, Vol. IV*. New York: Thomas Yoseloff, 1956.
Sandburg, Carl. *Abraham Lincoln: The War Years, Vols. I, II, III*. New York: Harcourt, Brace, & World, 1939.
Scharf, Thomas J. *History of the Confederate States Navy*. New York: Gramercy Books, 1996.
Snow, William P. *Lee and His Generals*. New York: Gramercy Books, 1867.
"Special Order 191." Wikipedia. http://en.wikipedia.org/wiki/Special_Order_191.

Swinton, William. "Army of the Potomac." *Report on the Conduct of the War, Vol. 1.* New York: W. S. Konecky Associates, 1995.
Taylor, Richard. *Deconstruction and Reconstruction.* Nashville: J. S. Sanders & Co., 1998.
Taylor, Walter H. *Four Years with General Lee.* New York: Bonanza Books, 1942.
Thomason, John. *Jeb Stuart.* New York: Charles Scribner's Sons, 1930.
Trimble, Isaac. "Confederate Veteran." Southern Historical Society Papers, Vol. 40.
Tucker, Glen. *High Tide at Gettysburg.* New York: Konecky & Konecky, 1958.
U.S. War Department. *The War of the Rebellion: A Compilation of the Official Records of the Union and Confederate Armies.* Series I, Vol. 5, Part I; Series I, Vol. 11, Part II; Series I, Vol. 11, Part III; Series I, Vol. 12, Part I; Series I, Vol. 12, Part II; Series I, Vol. 12, Part III; Series I, Vol. 16, Part I; Series I, Vol. 16, Part II; Series I, Vol. 19, Part I; Series I, Vol. 19, Part II; Series I, Vol. 21; Series I, Vol. 24, Part I; Series I, Vol. 24, Part III; Series I, Vol. 27, Part I; Series I, Vol. 30, Part II; Series I, Vol. 45, Part II; Series I, Vol. 46, Part I; Series I, Vol. 51, Part I Supplement. Harrisburg, PA: National Historical Society, 1971.
Wheeler, Gen. Joseph. "Bragg's Invasion of Kentucky." *Battles and Leaders of the Civil War, Vol. III.* New York: Thomas Yoseloff, 1956.
"Wood at Chickamauga." *New York Times,* December 25, 1881.

Index

Alabama 93
Alexandria 64, 65
Alexandria and Orange Railroad 64
Allegheny Mountains 25
Anderson, Governor 18
Anderson, Gen. R.H. 82, 87
Antietam Creek 78
Appalachian Mountains 25, 40, 76
Aquia 64
Army of East Tennessee 93
Army of Kentucky 93
Army of Northern Virginia 43, 48, 61, 66, 86, 100, 124, 126
Army of the Cumberland 138–139
Army of the Mississippi 67, 96
Army of the Ohio 88, 90
Army of the Potomac 13, 36, 40, 53, 62, 64, 67, 82, 85, 99, 104, 106, 124, 162, 169, 173
Army of the Tennessee 110
Army of Virginia 62–63, 64, 67, 69, 72
Articles of War 3, 190
Ashby, Gen. 30, 31
Ashland 40, 41, 43, 45–46
Atlanta 123, 138, 152, 156
Ayers, Gen. 176

Baker, Col. 15, 17–19, 21–24
Ball's Bluff 13–24
Baltimore and Ohio Railroad 86, 126
Baltimore Pike 132
Banks, Gen. N.P. 16, 27, 29, 30, 32, 34–35, 83, 107, 110
Battle of the Little Big Horn 181
Battle of the Rosebud 185
Battle of the Washita 190–191
Beauregard, P.G.T. 24, 91
Beaver Dam Creek 42–43, 45, 48
Benteen, Capt. Frederick 188, 191, 193–195
Big Black River 119
Big Horn 182, 184–187, 192–193
Big Round Top 132

Birney, Gen. 104
Bloody Knife 194
Blue Ridge Mountains 25
Bolton 119
Bond, Maj. 150
Boonsboro 78, 80, 83, 85
Bostick, Maj. 160
Bottom's Bridge 39
Bowen, Gen. 115
Boydton Plank Road 164, 170–171
Bragg, Gen. Braxton 88, 89, 91–97, 123, 138, 139, 141–143, 145–149, 151
Branch, Gen. 43, 44–45
Brannan, Gen. 149–151
Brent, Gen. 145–147
Brisbin, Maj. 185, 187, 195
Bristoe 68–69
Brownsville 119
Browsers Ford 135
Buckner 145–147
Buell, Gen. Don Carlos 88–94, 96–97, 138–139
Buford, Gen. 131
Bull Run 13, 19, 23, 25, 40, 54, 62–76, 80, 92, 98, 107, 125, 129, 151, 189
Bull Run Mountains 64–65
Burke 164
Burkittsville 83
Burnside, Gen. Ambrose 85, 98–101, 103–109
Burnside's Mud March 106

Calhoun, Lt. 188, 194
Carlisle 126, 136
Cedar Mountain 35
Cemetery Hill 128, 131–133
Central Railroad 27, 29, 40–41, 43–46
Centreville 15, 69, 73
Chamberlain's Creek 170
Chambliss, Gen. 134
Champion's Hill 119, 120, 122
Chancellorsville 54, 61, 124–125
Chattanooga 88, 93, 114, 138, 139, 146, 147, 153
Cheatham, Gen. 92, 153, 157, 159, 160

Chesapeake and Ohio Canal 14, 126
Chesapeake Bay 101
Cheyenne 181
Chickahominy River 39
Chickamauga 138–141, 143–145, 147, 149, 150–151
Chilton, R.H. 33, 43, 44, 48, 52, 54, 57, 59, 61, 86
ciphers 9, 11
City Point 166
Cleburne, Gen. 146–147, 157, 161
Clinton 115–122
codes 9–10
Cogswell, Col. 22
Colburn, Gen. A.V. 16
Cold Harbor 43, 162, 164
Colt Firearms 108
Columbia 153, 155, 157, 159–160
Conrad's Ferry 14, 19, 21
Cooke, Lt. 188, 194
Cooper, Gen. Samuel 113
Cooper's Gap 142
Corinth 88, 139
Corps of Observation 15, 16, 18, 20–21
Couch, Gen. D. 83
Cox, Gen. 159
Crampton's Gap 82–85
Crawford, Gen. 176–177
Crittenden 95, 138–139, 141, 147–149
Crook, Gen. 183–185
Culp's Hill 7, 128, 132–134
Cumberland River 88
Cumberland Valley 76, 126
Custer, Boston 188
Custer, Gen. George Armstrong 183–196
Custer, Tom 188

Dabney's Mills 171
Dakotas 181
Danville Railroad 164, 166
Danville Road 166
Davies, Gen. 170–171
Davis, Jefferson 4, 44, 53–54, 60, 92, 113–114, 116, 123, 152, 156

Index

Davis's Crossroads 142, 145
Dawkins Branch 71–75
Department of Missouri 174, 178
Department of the Gulf 110
Department of the Ohio 90, 106
Department of the Pacific 90
Devens, Col. 20–23
Devin, Gen. 170–171
Dillon 117–119
Dinwiddie Court House 166–171, 177–178
Doubleday, Gen. 131
Dranesville 16, 17
Duck River 153, 157, 159
Dug Gap 142, 145–146

Edwards Depot 117, 119
Edwards Ferry 14, 19–21
Edwards Station 115–117, 122
Elizabeth City 98
Elliott, Maj. 190–191
Episcopal Church 92
Evans, Col. Nathan G. 16–17, 19, 24
Ewell, Gen. R.S. 7, 27–30, 32–34, 124–126, 128–131, 133–134, 136

Falconer, Kinloch 142, 144, 147
Far West 183–185, 187, 192
Farragut, Adm. 110
Fifth Corps 67, 166, 171, 173
Fifth N.Y. Cavalry 29
Fifth Virginia 44
First Bull Run 125, 151
First Maryland Infantry 29
Fisher's Hill 35
Five Forks 162–163, 167–177
Flournoy, Col. 32–34
Ford Road 167–168, 175–176
Ford's Crossing 167–168
Ford's Theater 107
Forrest, Gen. Nathaniel Bedford 21, 91, 157, 160
Forsyth, J.W. 177
Fort Donelson 88
Fort Ellis 183
Fort Fetterman 183
Fort Fisher 189
Fort Henry 88
Fort Leavenworth 190
Fort Lincoln 183
Fort Macon 98
Fort Monroe 36–38, 53, 58
Fort Stedman 165, 168
Fourth Corps 159
Fox's Gap 80, 85
Frankfort 93, 94, 96
Franklin, Gen. William Buel 38–39, 80–87, 98–101, 103–109
Franklin, Tennessee 153–154, 157, 159–161, 164
Frederick 76, 77, 80, 82, 85, 136, 137

Fredericksburg 40, 54, 64, 67, 98, 100–103, 105, 107, 125
Front Royal 27, 29, 30

Gaines Mill 36, 47, 49–51, 55, 67
Gainesville 69–71, 73, 75
Garnett, Gen. 33
Gatling Gun 183
General Order 75, 46–48
George Washington College 35
Gettysburg 5, 7, 54, 113, 120, 124–125, 128–133, 135–137, 156, 161, 172–173
Gibbon, Col. John 183–187, 192, 194–196
Gibbon, Gen. John 172
Gist, Gen. 116, 118
Gloucester 37–38
Godfrey, Gen. 187
Goldsborough 166
Gordon's Mills 142–143, 145–146, 149
Goreman, Gen. 15, 19–20
Grand Gulf 114, 118
Grant, Gen. Ulyssees S. 5–6, 8, 31, 54, 88, 90, 92–93, 99, 106–107, 110, 112, 114–115, 117, 119–121, 129, 141, 152, 162, 164–166, 168–172, 174, 177–178, 189, 194–196
Gravelly Run 169, 171
Gregg, Gen. 116–118
Griffin, Gen. 73, 75, 177
Grover, Col. Ira 128, 133–134
Groveton 70, 72

Hagerstown 78, 83, 86, 87
Half Sink 41, 43–44, 48
Halleck, Gen. Henry 35, 85, 88, 141, 174
Hamilton's Crossing 102–105, 108
Hampton, Gen. Wade 44, 82
Hancock, Gen. 128, 132
Hanover 136
Hard Times 112, 114
Hardee, Gen. 94–95, 97, 141
Harpers Ferry 6, 9, 13, 15, 27, 35, 61, 77–87, 100, 107
Harpeth River 153, 160
Harrisburg 76, 126
Harrisonburg 25
Harrison's Island 17, 20–21
Harrodsville 94
Hartford 108
Hatcher's Run 168, 176
Hawes, Governor 93–94
Haynes Bluff 120
Heintzelman, Gen. 38–39, 69
Heth, Gen. 131
Hill, Gen. A.P. 38, 41, 43–45, 47–48, 55–56, 58, 87, 125, 131, 172
Hill, Gen. D.H. 6, 9, 38, 41, 43, 45–46, 48, 50, 55–57, 59,

61–62, 78, 83–84, 87, 141–143, 145, 148–149
Hindman, Gen. 142–149
Hood, Gen. J.B. 123, 152–161, 164
Hooker, Gen. J. 101, 103–104, 108, 124
Howard, Gen. 131
Howe, Lt. 21–22
Huger, Gen. 38, 41, 43–44, 55, 57–59
Humphreys, Gen. 165–166, 168–170
Hundley's Corner 46
Huntsville 93

Interior Department 181

Jackson, Gen. T.J. "Stonewall" 6, 8, 19, 25, 27–35, 40–41, 43–48, 55, 57–59, 62, 64–68, 70, 72, 75, 77–80, 82–83, 85–87, 90, 101, 125, 129, 141, 172
Jackson, Mississippi 113–118, 121–122
James River 39, 50, 51, 57
Jefferson 80, 82–83
Johnson, Gen. 133–134, 153, 156, 160
Johnston, Gen. Albert Sydney 6, 88, 91,
Johnston, Gen. Joseph E. 27, 38–39, 53–54, 113–123, 152, 156, 164
Joint Committee 23, 105

Kanawa Division 67
Kelly, Captain 150
Kelly, Mark 188
Kenly, Col. 29
Kentucky 89–90, 92–94, 97, 138, 152
Keyes, Gen. E. 38–39
King, Gen. R. 68–71
Knoxville 93, 106, 138

Lafayette, Georgia 80, 139, 141–142, 146–149
Lakota War 178, 181, 196
Landers, Gen. 15, 19, 22
Laurel Cemetery 123
Lawton, Gen. 87
Lee, Col. 20–22
Lee, Gen. Fitzhugh 134, 176–177
Lee, Gen. Robert E. 6, 7, 19, 33, 39, 40–41, 43–51, 53–55, 58–59, 61–62, 64–67, 76–77, 80, 82, 84, 86, 90, 92, 99, 101, 105, 113–114, 124–126, 129, 131, 133, 136, 142–143, 145–146, 149, 152, 156–157, 159, 161–162, 164–166, 168–172, 176, 189

Index

Lee, Gen. S.D. 155
Lee and Gordon's Mills 142–143, 145–146, 149
Leesburg 14, 16, 17, 19–20, 76, 86, 126
Lexington 34, 93, 96
Liddel's Brigade 96
Lincoln, President Abraham 18–19, 27, 36, 39, 40, 76, 98, 101, 106–107, 124, 162, 172, 174, 183
Little Big Horn 181, 185–187, 189, 191–193, 195–196
Little Round Top 132, 173
Locke, Lt. Col. 71, 72
Longstreet, Gen. J. 6, 38, 41, 43, 45–46, 48, 55–58, 64–67, 71, 75, 78, 83, 87, 101, 124–125, 137, 139, 141, 149, 151
Lookout Mountain 142, 147, 148
Loring, Gen. 118
Louisiana 91, 107, 110, 113, 120
Louisiana State University 91
Louisville 93
Luray 26–27, 29

MacKenzie, Gen. 171, 175
Magruder, Gen. John B. 38, 41, 43, 52–55, 57–60
Malvern Hill 7, 49–60, 67, 161
Manassas 62, 64–66
Manassas Junction 64, 68, 69–71, 73, 75
Mansfield 107
Martin, Gen. 142, 145
Martinsburg 27, 30–31, 77–78, 82, 86
Martinsburg Turnpike 31–34
Maryland 5, 13–15, 18, 21, 29, 32, 61, 76, 126, 136
Maryland Campaign 76, 79, 81, 86, 98
Maryland Heights 78, 83, 86–87
Mason, Maj. A.P. 159
Massanutten Mountain 25, 29
Maxey, Gen. 116, 118–119
McCall, Gen. 16–17
McClellan, George B. 5–6, 13, 15–17, 20, 25, 27, 31, 35–40, 47, 49–51, 53–54, 57–58, 60–64, 67, 76–78, 80–85, 87, 90, 98–101, 107, 124, 129, 139, 141, 172
McCook, Gen. 95, 138–139, 141, 148
McDowell, Gen. Irvin 27, 29, 39–41, 42, 65, 67–72
McLaws, Gen. L. 78–87
McLemore's Cove 141–144, 147–148
McMahon, Col. 103
McPherson, Gen. 118–119
Meade, Gen. George G. 104–105, 107, 113, 162, 165, 169

Mechanicsville 36–38, 41–43, 45–49, 55, 62
Memphis-Charleston RR 88
Mexican War 3–5, 17–19, 53–54, 77, 90–92, 98, 113, 128, 139, 172
Middleton 30
Middletown 83, 86–87
Miles, Col. Dixon 77–78, 82–84
Missouri 153
Missouri River 183, 191
Mobile-Ohio RR 88
Montana 181, 186
Morell, Gen. 71, 73
Morgans 145–146
Munfordville 93

Nashville 90, 93, 138, 153, 154, 156, 160–161, 164
National Road 77
Negley, Gen. 142, 146, 150
New Bridge 43, 55
New Market 26–29
New York Times 139, 188, 191
Newton 30
Nineteenth Corps 107
Ninth Corps 67, 106
Ninth Virginia Cavalry 44
Nocquet, Maj. 147–148
North Carolina 44, 98, 166

Ohio River 89

Parker, Gen. John 104
Parrott cannon 29
Pegram, Gen. 149
Pemberton, Gen. J. 113–123
Pendleton, Lt. "Sandy" 31–35, 54
Peninsular Campaign 36, 62, 129
Perryville 88–90, 93–97, 138, 149
Petersburg 106, 162, 164, 166, 169
Philbrick, Captain 19–21
Pickett, Gen. G. 161, 168–173, 175–178
Pigeon Roost Mountain 142, 145
Pleasant Valley 80, 82–84
Pleasants, Col. Henry 106
Point of Rocks 15, 16
Pole Green Church 43, 46
Polk, Gen. Leonidas 91–92, 94–97, 141–142, 147–149
Poolesville 15–17, 20
Pope, Gen. John 62, 64, 65, 67–76
Port Gibson 115, 117
Port Hudson 110, 112–114, 115–116, 118, 121
Porter, Adm. 110, 112
Porter, Gen. Fitz-John 38, 39, 41–44, 47, 65, 67–75
Porter, Horace 168

Potomac River 13–15, 25–27, 30, 32, 63, 76–79, 81–83, 85–87, 99, 126, 134–137
Powder River 182–185

Rappahannock River 40, 63–65, 100–102, 108
Rappahannock Station 64, 100
Raymond 115, 117–119
Rectors Crossroads 134, 136
Red River 110–111
Red River Campaign 107
Reed, Henry 188, 194
Reno, Gen. Jesse 69, 172
Reno, Maj. Marcus 184–186, 188, 190–191, 193–195
Reserve Corps 67
Reynolds, Gen. John 72, 103, 131, 150–151
Rhode Island 107
Richmond 25, 27, 29, 36–41, 43, 47, 49–51, 53–55, 58, 60–64, 66, 93, 100, 102, 104, 106, 115, 162–166
Ricketts, Gen. 68–71
Roanoke Island 98
Robertson, Gen. 134
Rockville Pike 135
Rohrersville 80, 82–83, 85
Rome, Georgia 139
Rosebud River 182, 184–188, 190, 192–193, 195
Rosecrans, Gen. William Starke 138–141, 149–151
Rosser, Gen. 171, 176
Rutherford's Creek 153, 157

St. Louis Democrat 191
Schofield, Gen. J.M. 153–157, 159–161, 164
Scott, Gen. W. 5
Second Battle of Bull Run 62, 64–66, 80, 98, 129
Second Cavalry 6, 19, 187
Second Corps 125–126, 128, 137, 173
Seminole War 53, 90, 91
Seven Days Battle 47, 49, 54–55, 57, 60, 62
Seven Pines 39
Seventh Brigade 19
Seventh Cavalry 183–185, 187–188, 190, 191
Seventh Indiana 133
Sharpsburg 78–80, 83–86
Shenandoah River 25, 87, 136
Shenandoah Valley 15, 25–27, 40, 76, 126, 174
Shenandoah Valley Army 164
Shepherdstown 136–137
Sheridan, Gen. P. 129, 164–172, 174–179, 184, 189–190
Sherman, Gen. William T. 6, 91, 116, 129, 152, 164, 166, 174
Sigel, Gen. F. 69

Sioux 181
Sitting Bull 181
Sixth Corps 82, 85, 100
Sixth Virginia Cavalry 32
Smith, Capt. E.W. 187
Smith, Col. 16
Smith, Gen. Edmund Kirby 92–97, 103, 170
Snyders Mill 120
South Mountain 80, 82–84, 100, 107
Southside Railroad 164, 166–167, 169, 176
Special Order 191 78–80, 82, 86, 87
Spring Hill 152–155, 157–161
Springfield 45, 188
Stanley, Gen. 153–155, 157, 159–160
Stanton, Secretary of War E.M. 58, 141
Staunton 29
Stephenson's Depot 32
Steuart, Gen. 30–34
Stevens, Maj. 44
Stevens Gap 142–143, 146–147
Stewart, Gen. 153, 157–160
Stone, Gen. Charles P. 15–23
Stoneman, Gen. George 38, 172
Stone's River 141
Strasburg 27–30
Stuart, Gen. J.E.B. 19, 38, 44–46, 48, 66, 82, 87, 101, 124–126, 128–130, 134–137
Sturgis, Col. 188
Sturgis, Gen. Samuel 73, 75, 172
Sudley Springs Road 69, 71, 73
Sumner, Gen. 38–39, 83, 101, 103–105, 108–109
Sumner, Sen. 18
Susquehanna River 126
Sykes, Gen. 71, 73, 83

Taylor, Maj. 86
Taylor, Gen. Richard 130
Taylor, Walter 133
Taylor, Gen. Zachary 5
Tennessee 88–89, 93, 97, 106, 114–115, 138–139, 152–154
Tennessee River 88, 138, 142
Tenth Virginia Cavalry 44
Terry, Gen. 183–187, 189, 192, 194–195
Third Corps 125
Third Virginia Cavalry 44
Thomas, Gen. G. 138–139, 141–142, 146–151, 153, 161, 164
Thompson's Station 153, 157, 160
Thoroughfare Gap 64–65, 68, 70–71
Tongue River 182, 184–187, 195
Totopotomoy Creek 46
travois 183
Trimble, Gen. Isaac 46–47, 133
Tullock's Creek 187
Tupelo 88, 91
Turner's Gap 77, 80, 82–85
Twelfth Corps 134
Twenty-ninth Pennsylvania Infantry 29
Twenty-third Corps 159

Valley Army 40
Valley Campaign 25, 34, 129
Valley District 25, 87
Vancleve, Gen. 150
Varnum, Lt. 191
Vicksburg 110–116, 118–122
Virginia 13–14, 16, 20–23, 25, 33, 36, 40, 53, 62, 76–77, 100, 113, 123–124, 126, 130, 134, 136, 138, 152, 168, 169, 173
Virginia Central Railroad 27, 29, 41, 44, 46
Virginia Military Institute 35
Virginia Theological Seminary 92

Wadsworth, Gen. 128, 132–133
Walker, Gen. 6, 78–80, 85–87, 116, 142
War Department 27, 57, 61, 64, 67, 99, 114–116, 149, 181, 183
Warren, Gen. Gouverneur 165, 166, 168–173, 177–178
Warrenton 64, 136
Warrenton Pike 68–70, 72–73
Washington 13, 15–16, 19, 23, 27, 36, 39, 40, 49, 58, 62–64, 66–67, 76, 99–101, 126, 135–136, 174
Washington, Gen. George 5
Waterloo 106, 141
Weldon Railroad 164
West Point 4–6, 17–29, 23, 32, 53, 54, 77, 90–92, 98, 113, 128–129, 139, 141, 155–156, 172–174, 189–191
West Virginia 67, 139, 140
Wheeler, Gen. J. 94–95, 97, 149
White, Gen. 77
White House 37–40, 43, 47, 49
White Oak Bridge 51, 56–57
White Oak Road 165, 167–171, 175–176, 178
White Oak Swamp 51, 55–57
Whiting, Gen. 46
Wilcox, Gen. Cadmus 172
Williamsport 30, 83
Winchester 15, 25–26, 28–31, 34–35, 86
Wood, Gen. 150–151
Woodbury's Bridge 47
Wool, Gen. 5

Yellowstone National Park 179
Yellowstone River 182–185, 187, 192
York, Pennsylvania 126, 134–137
York, Virginia 39
York Peninsula 25, 34, 49, 53, 58, 62
York River 36–38, 43

www.ingramcontent.com/pod-product-compliance
Ingram Content Group UK Ltd.
Pitfield, Milton Keynes, MK11 3LW, UK
UKHW050132170426
5217IPUK00052BA/1830